Synthesis and Characterization of Nanomaterials

Synthesis and Characterization of Nanomaterials

Editor

Ioannis Kartsonakis

MDPI • Basel • Beijing • Wuhan • Barcelona • Belgrade • Manchester • Tokyo • Cluj • Tianjin

Editor
Ioannis Kartsonakis
School of Chemical Engineering
National Technical University of
Athens
Zographos Attikis
Greece

Editorial Office
MDPI
St. Alban-Anlage 66
4052 Basel, Switzerland

This is a reprint of articles from the Special Issue published online in the open access journal *Fibers* (ISSN 2079-6439) (available at: www.mdpi.com/journal/fibers/special_issues/Synthesis_Characterization_Nanomaterials).

For citation purposes, cite each article independently as indicated on the article page online and as indicated below:

LastName, A.A.; LastName, B.B.; LastName, C.C. Article Title. *Journal Name* **Year**, *Volume Number*, Page Range.

ISBN 978-3-0365-2943-1 (Hbk)
ISBN 978-3-0365-2942-4 (PDF)

Contents

About the Editor

Ioannis Kartsonakis

Ioannis Kartsonakis is a Chemist from the National and Kapodistrian University of Athens (NKUA), Greece. His masters degree is in Polymer Science (NKUA) with subject on synthesis and characterization of standard polymers. He has a Ph.D. in Chemistry (NKUA in collaboration with the National Centre for Scientific Research "Demokritos") with subject on Corrosion protection of metal alloys, synthesis and characterization via combination of the Sol-Gel method, electrochemistry and polymer science of hybrid organic-inorganic coatings, as well as micro and nanocontainers. His scientific subjects are focused on synthesis and characterization of materials, such as nanocomposites, nanocontainers, coatings (polymeric, inorganic, hybrid organic-inorganic) and study of corrosion processes on materials. He has experience in many European and National projects related to: Recycling of textile and plastic materials; Concrete Reinforcement; Self-healing, Antifouling, Anticorrosion, coatings and materials. His work is focused on management of R&D projects; synthesis and characterization of hybrid inorganic/ organic nanomaterials and composites; corrosion issues consultancy.

Preface to "Synthesis and Characterization of Nanomaterials"

Nanomaterial is defined as a natural, incidental, or manufactured material containing particles, in an unbound state or as an aggregate or as an agglomerate and where, for 50 % or more of the particles in the number size distribution, one or more external dimensions is in the size range 1 nm to 100 nm. In specific cases, and where warranted by concerns for the environment, health, safety, or competitiveness, the number size distribution threshold of 50 % may be replaced by a threshold between 1 and 50 %. The classification of the nanomaterials can be based on the type of materials they consist of, such as metallic nanomaterials, carbon-based nanomaterials, polymeric nanomaterials, ceramic-based nanomaterials, and biomolecules derived nanomaterials. Nanomaterials not only differ from the corresponding bulk materials in morphological properties but they can also demonstrate different physico-chemical characteristics. Due to this fact, changes in chemical properties, (photo-) catalytic activities, reactivity and energetic properties of the nanomaterials can be caused, making them potential candidates for excellent performance in many applications.

The introduction of innovative functions in products and technologies can be possible through nanomaterials. New products and devices that present improved efficiency than conventional bulk materials, have been fabricated based on nanomaterials. Therefore, manufactured nanomaterials are regarded as key components of innovations in various fields with high potential impact, such as energy generation and storage, electronics, photonics, diagnostics, integrated sensors, semiconductors, foods, textiles, structural materials, sunscreens, cosmetics, and coatings or drug delivery systems, and medical imaging equipment. Widespread use of nanomaterials raises concerns about their safety for humans and the environment, possibly limiting the impact of the nanotechnology-based innovation. The development of safe nanomaterials has to result in a safe as well as functional material or product. Its safe use, and disposal at the end of its life cycle must be taken into account too. Responsibility for the safe handling of synthetic nanomaterials therefore rests with the manufacturer and importer.

In order to update the field of development, synthesis, structure-activity relationships and future applications of nano-materials, a Special Issue entitled "Synthesis and Characterization of Nanomaterials"has been introduced. This book gathers and reviews the collection of five contributions (four articles and one review), with authors from Europe and America accepted for publication in the aforementioned Special Issue of *Fibers*. This book aims to attract all researchers working in the research field related to the development, synthesis, structure-activity relationships, and future applications of nanomaterials.

Ioannis Kartsonakis
Editor

fibers

MDPI

Article

New Textile for Personal Protective Equipment—Plasma Chitosan/Silver Nanoparticles Nylon Fabric

Cláudia M. Botelho [1,*], Margarida M. Fernandes [1,2], Jefferson M. Souza [3], Nicolina Dias [1], Ana M. Sousa [1], José A. Teixeira [1], Raul Fangueiro [4] and Andrea Zille [4]

1 CEB—Centre of Biological Engineering, Universidade do Minho, 4710-057 Braga, Portugal; margaridafernandes@fisica.uminho.pt (M.M.F.); nidias@deb.uminho.pt (N.D.); anamargaridasousa@deb.uminho.pt (A.M.S.); jateixeira@deb.uminho.pt (J.A.T.)
2 Centro/Departamento de Física, Universidade do Minho, 4710-057 Braga, Portugal
3 CBMDE, Design and Styling, Federal University of Piauí, Teresina, PI 64049-550, Brazil; jefferson@ufpi.edu.br
4 2C2T—Centro de Ciência e Tecnologia Têxtil, Universidade do Minho, 4800-058 Guimarães, Portugal; rfangueiro@civil.uminho.pt (R.F.); d3742@det.uminho.pt (A.Z.)
* Correspondence: claudiabotelho@deb.uminho.pt; Tel.: +35-19-6636-5977

Abstract: Fabric structures are prone to contamination with microorganisms, as their morphology and ability to retain moisture creates a proper environment for their growth. In this work, a novel, easily processed and cheap coating for a nylon fabric with antimicrobial characteristics was developed. After plasma treatment, made to render the fabric surface more reactive sites, the fabric was impregnated with chitosan and silver nanoparticles by simply dipping it into a mixture of different concentrations of both components. Silver nanoparticles were previously synthesized using the Lee–Meisel method, and their successful obtention was proven by UV–Vis, showing the presence of the surface plasmon resonance band at 410 nm. Nanoparticles with 25 nm average diameter observed by STEM were stable, mainly in the presence of chitosan, which acted as a surfactant for silver nanoparticles, avoiding their aggregation. The impregnated fabric possessed bactericidal activity higher for Gram-positive *Staphylococcus aureus* than for Gram-negative *Pseudomonas aeruginosa* bacteria for all combinations. The percentage of live *S. aureus* and *P. aeruginosa* CFU was reduced to less than 20% and 60%, respectively, when exposed to each of the coating combinations. The effect was more pronounced when both chitosan and silver were present in the coating, suggesting an effective synergy between these components. After a washing process, the antimicrobial effect was highly reduced, suggesting that the coating is unstable after washing, being almost completely removed from the fabric. Nevertheless, the new-coated fabric can be successfully used in single-use face masks. To our knowledge, the coating of nylon fabrics intended for face-mask material with both agents has never been reported.

Keywords: fabric; infection; chitosan; silver nanoparticles; antimicrobial; facemask

Citation: Botelho, C.M.; Fernandes, M.M.; Souza, J.M.; Dias, N.; Sousa, A.M.; Teixeira, J.A.; Fangueiro, R.; Zille, A. New Textile for Personal Protective Equipment—Plasma Chitosan/Silver Nanoparticles Nylon Fabric. *Fibers* **2021**, 9, 3. https://doi.org/10.3390/fib9010003

Received: 4 November 2020
Accepted: 1 December 2020
Published: 6 January 2021

1. Introduction

Considering the actual pandemic situation that we are living and the increasing use of masks by the population, it is of paramount importance to develop novel antimicrobial textiles that could be used either in a medical context, for hospitalized patients and hospital staff, or for generalized use as face mask. Fabrics have an ideal environment for the growth of microorganisms. They are porous, retain water and, as they are close to human body, their normal temperature is approximately 37 °C. The constant use of facial masks and protective clothes, if not used properly, increases the risk of infection. A fabric with antimicrobial properties is, in fact, essential to overcome this problem.

The antimicrobial properties of silver have been known for many years. Silver is currently used in several medical devices, like wound dressings, coating of surgical instruments and prostheses [1]. With the advances of technology, silver nanoparticles (AgNPs)

have been widely prepared and demonstrated to possess high antimicrobial potential. They are now considered a good alternative to antibiotics, although their use for medical applications can be hindered due to its high toxicity to the patient and the environment [2]. It is known that AgNPs can enter the human body by different ways and accumulate in different organs, but most important, they can cross the blood–brain barrier and reach the brain [1]. Thus, there is the need to decrease the AgNPs' toxicity but keep their antimicrobial properties. One approach is the incorporation of the AgNPs in a biocompatible polymer like chitosan, which has intrinsic antimicrobial properties, creating a synergy between the two materials [3,4]. An interesting fact is that AgNPs nanoparticles have demonstrated an efficient inhibitory activity against human immunodeficiency virus (HIV-1) [5,6], hepatitis B virus (HBV) [7], respiratory syncytial virus herpes simples virus type 1 [8,9], monkeypox virus plaque formation [10] and influenza H1N1 virus [11]. A very recent study claims that silver may be effective on the treatment and prevention of COVID-19 and SARS-COV-2 [12].

Plasma-assisted polymerization and sputtering is one of the most used techniques for nanomaterials' deposition at an industrial scale [13,14]. However, this technique presents significant limitations like low deposition rate, sample size, batch operation, limited electrodes spacing and expensive vacuum systems [15]. Atmospheric pressure plasma can be considered an alternative to plasma-assisted polymerization and sputtering, as it is a cost-competitive method when compared to wet chemical treatments and low-pressure plasma, allowing continuous and uniform processing of fibers' surfaces [16]. Double Dielectric Barrier (DBD) plasma is considered the most effective cold atmospheric plasma source, and it has been widely applied for surface activation and deposition of thin functional coatings [17,18].

Therefore, in this work, a DBD plasma pretreated nylon fabric was functionalized by a simple dip coating method with a mixture of different concentrations of AgNPs and chitosan. The characteristics of the textile were analyzed, as well as the synergistic effect of the active agents. Silver-based nylon fabric has long been known to act as an antimicrobial agent for wound therapy, helping the mitigation of wound infection and inflammation [19]. The synergistic effect of silver and chitosan has also been proven to be effective for preventing the agglomeration of AgNPs and improve its biocompatibility [3,20]. To our knowledge, a nylon coating intended for face masks material with both agents has never been reported.

2. Materials and Methods

2.1. Materials

Chitosan (ChitoClear hq95-43000) was acquired from Primex (Siglufjordur, Iceland), and polyamide taffeta with 52 warp and 32 weft yarns, with an areal density 100 g m^{-2}, was acquired from Lemar (Guimarães, Portual). All the other reagents were purchased from Sigma-Aldrich (St. Louis, Missouri, EUA) and used without further purification.

2.2. Nanoparticles Synthesis

A 0.02 gL^{-1} colloidal dispersion of AgNPs was prepared, using the modified method described by Lee and Meisel [21]. All solutions were prepared in distilled water, and the dispersion was subjected to a vigorous mixing. A volume of 10 mL of 1% trisodium citrate ($Na_3C_6H_5O_7$), as reducing agent, was added drop-by-drop to 100 mL of 1 mM silver nitrate ($AgNO_3$), which was previous heated to boiling point in a 250 mL flask, reaching a final concentration of 3.8 mM. The pH of the reaction was adjusted to 7.7 by adding NaOH at the end of the fast nucleation stage, to stop the reaction. The solution was heated again, to boiling temperature, until color's change was evident (pale yellow). At this stage, the solution was cooled to room temperature, while mixing. The resulting colloidal solution presented high stability, i.e., the aggregation of the NPs was very low for 2 months, when stored at 4 °C.

2.3. Preparation of Chitosan Solution

Chitosan was prepared in a concentration of 0.5 wt%. Chitosan (0.5 g) was dissolved in 2 mL of acetic acid (glacial), to completely dissolve the chitosan, and distilled water was further added, until we reached 100 mL of solution, and stirred at 300 rpm for 30 min.

2.4. Nylon Fabric Plasma Treatment

To improve the adhesion of the coating, plasma treatment was applied prior to the application of the antimicrobial agents. The Double Dielectric Barrier plasma was applied in a semi-industrial prototype machine (Softal Electronics GmbH/University of Minho, Guimarães, Portugal), at room temperature and atmospheric pressure. A metal electrode coated with ceramic and counter electrodes coated with silicon with an effective width of 50 cm and a gap distance of 3 mm were used to induce a 10 kV high voltage discharge and a 40 Hz frequency. Some variations may occur, including the supplied discharge power and the speed, which can reach a maximum of 1.5 kW and a speed of 60 m min^{-1}. The optimized parameters used were 1 kW of power and a velocity of 4 m min^{-1} with 5 passages corresponding to a dosage of 2.5 kWmin m^{-2}, as previously reported [22]. A pre-washed sample is simply passed through the plasma lamina 5 times, in both sides, to obtain the desired dosage. A significant increase in the oxygen content due to the incorporation of oxygen atoms onto the fabric surface is usually observed after plasma treatment. Plasma etching may provoke some scission in the C-H, C-O, C-N, C-C and N-H bonds present in the fibers, leading to the formation of reactive O^-, N, N^+, O, OH^- and O^3 species [23]. In addition, plasma treatment also induces notable morphological alterations on the fibers' surface, augmenting the roughness and thus increasing surface energy, wettability and adhesion [24].

2.5. Functionalization of Nylon Fabric with Antimicrobial Agents

Nylon fabrics were coated with chitosan, AgNPs and a mixture of both, using different concentrations of mixed solutions, as described in Figure 1. Dilutions were made starting with the mother solution of chitosan (CH) and AgNPs (Ag) with the following percentage ratio of CH/Ag: 100:0, 0:100, 60:40, 70:03, 80:20 and 90:10 (Figure 1A). The coating solution was applied to the polyamide fabric by a simple dip-coating technique. The fabric was dipped in each solution for 5 min, at room temperature (Figure 1B). The coated material was incubated in an oven, at 50 °C, for 20 min.

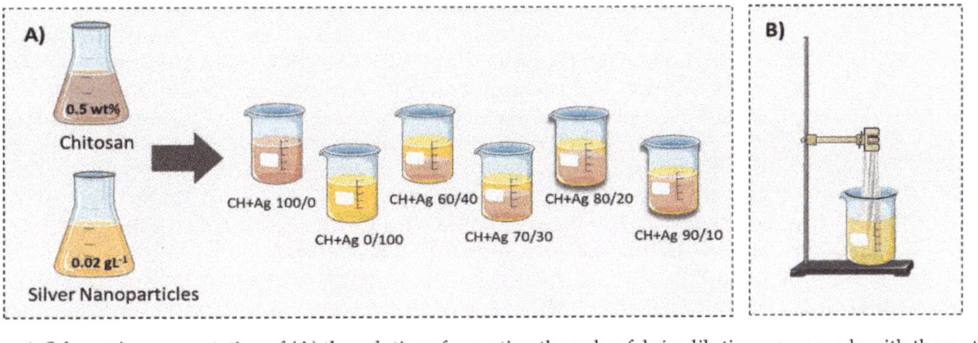

Figure 1. Schematic representation of (**A**) the solutions for coating the nylon fabric; dilutions were made with the mother solution of chitosan (CH) and silver nanoparticles (Ag), and (**B**) dip-coating method used for the coating. The dilutions are represented as the percentage of chitosan and silver (CH/Ag).

2.6. Scanning Electron Microscopic (SEM) Analysis

The morphological characterization of AgNPs and the combination of both AgNPs with chitosan were visualized by SEM, which was also used to assess the mean particle average of the obtained NPs. Further, the functionalized nylon fabrics with both antimicrobial agents before and after application of plasma treatment were also visualized with SEM. An ultra-high resolution Field Emission Gun–Scanning Electron Microscopy (FEG–SEM), NOVA 200 Nano SEM, FEI Company (Hillsboro, Oregon, EUA) was used. Secondary electron and backscattering electron images were performed with an acceleration voltage of 5 and 15 kV, respectively. Samples were previously coated with an Au-Pd film (80–20 weight %) in a high-resolution sputter coater, 208HR Cressington Company (Watford, England), coupled to an MTM-20 Cressington High-Resolution Thickness Controller, to reduce the possibility of oxidation during processing. A turbo pump was used to obtain a higher (and cleaner) vacuum environment. To remove air and increase sputter efficiency, pure argon gas was backfilled into the chamber. When high-magnification images are required, Au alone is not recommended as a sputter coating. The recommended metal coatings for general research purpose is gold/palladium sputtered alloy (80/20), due to its having a smaller grain size than Au. The Pd L-series X-ray lines at 2.84 keV do not overlap important lines from other elements; thus, no additional interference with X-ray microanalysis would be expected besides the one mentioned for Au.

The prepared silver colloid suspension was released in copper grids with carbon film 400 meshes, 3 mm diameter for STEM nanoparticle analysis. The mean size was extrapolated by measuring the average diameter based in 50 observations.

2.7. *UV–Vis Reflectance and Adsorption Measures*

UV–visible absorption intensity employing diffuse reflectance spectrophotometer (Spectraflash 600, Datacolor, Lawrenceville, NJ, USA) at standard illuminant D65 (LAV/Spec. Excl., d/8, D65/10°) was used to monitor the changes associated with modification of nylon fabric with silver nanoparticles in the wavelength range of 200–800 nm, at room temperature. The results were expressed as the values of reflectance percentage at the peak of maximum adsorption of the AgNPs. The peak of the maximum adsorption of the AgNPs in the aqueous colloidal solution, determined at 410 nm and characteristic of silver surface plasmon resonance (SPR), was measured in a UV–Vis spectrophotometer (Shimadzu, Kyoto, Japan, UV-1800) with quartz cuvettes (Figure 2C).

2.8. *X-ray Photoelectron Spectroscopy (XPS)*

XPS was performed in a Kratos AXIS Ultra HAS (Kratos Analytical Limited, Manchester, UK), using a VISION software (Vision 2, Kratos Analytical Limited, Manchester, UK) for data acquisition. The acquired data were analyzed by using the CASAXPS software (casaXPS 2.3.22. Casa Software Ltd., Teignmouth, UK). A monochromatic Al Kα X-ray source (1486.7 eV), operating at 15 kV (150 W), in FAT (Fixed Analyzer Transmission) mode, was used, with a pass energy of 40 eV for regions ROI and 80 eV for survey. For data acquisition, a pressure lower than 10-6 Pa was applied, and a charge neutralization system was used. To obtain the s C1s spectral component (C-C. C-H) with a binding energy of 285 eV, the spectra were corrected, and a high-resolution spectrum of Ag was collected on a $\approx$1 mm^2 piece of polyamide fabric coated with only silver (without chitosan). A process to obtain a peak with equal FWHM (Full Width at Half Maximum) to the main peak was applied, inducing an error of approximately $\pm$0.1 eV.

2.9. *Antimicrobial Assays*

The antimicrobial activity was determined by using the standard shake flask method (ASTM-E2149-01), with some modifications. *Staphylococcus aureus* (ATCC25923) (*S. aureus*) and *Pseudomonas aeruginosa* (PAO1) (*P. aeruginosa*) inoculum were prepared by using a single colony from the corresponding bacterial. The culture was grown overnight in sterile nutrient broth (NB, Sharlab, Spain), at 37 °C and 230 rpm. The material was incubated with 5 mL of bacterial suspension (3E8 cell/mL in NaCl 0.9% pH 6.5), at 37 °C and 100 rpm, for 2 h. After the suspensions were withdrawn, the bacteria viability was determined by flow cytometry, as described below.

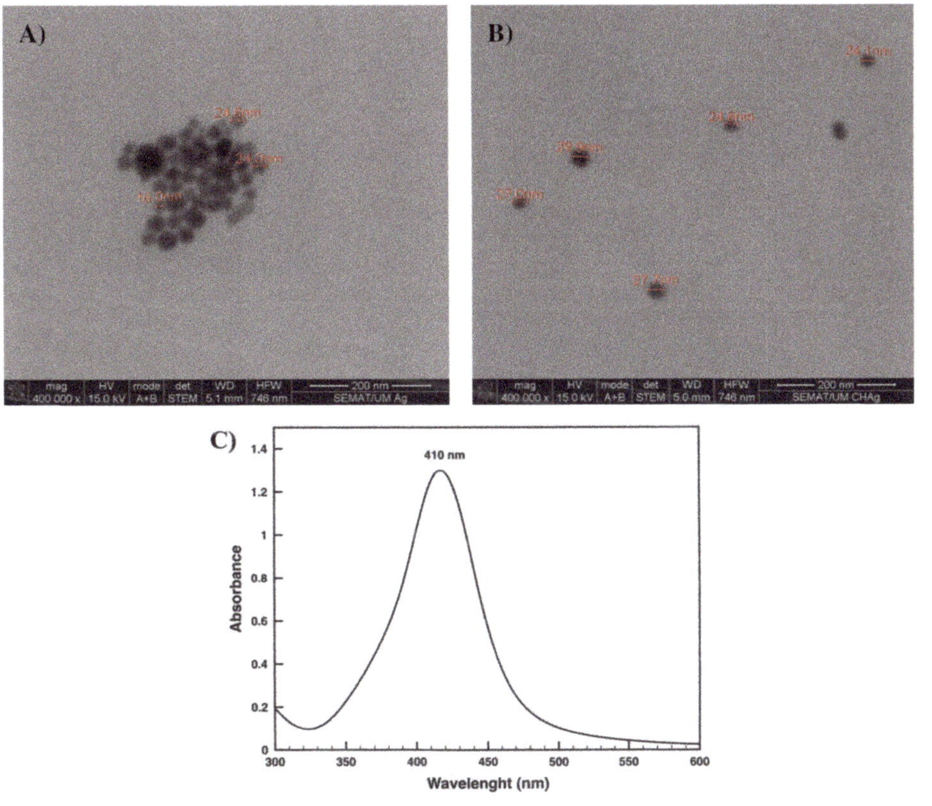

Figure 2. STEM images of the AgNPs aqueous colloidal solution depicting the synthesized silver nanoparticles in the absence (**A**) and the presence (**B**) of chitosan and (**C**) UV–Vis spectra of the AgNPs aqueous colloidal solution.

2.10. Flow Cytometry

Viability Staining Protocol: To assess bacterial viability, SYTO-BC (Molecular Probes Eugene, OR) and propidium iodide (PI) (Sigma-Aldrich, St. Louis, MO, USA) were used, according to manufacturer's instructions.

A volume of 2.5 µL and 10 µL of the SYTO-BC (50 µM stock solution) mixed with 5 µL of PI (1.5 mM stock solution) was added to 250 µL of bacterial suspension of *S. aureus* and *P. aeruginosa*, respectively. Samples were analyzed after at least 15 min of incubation at room temperature. Cell debris and other particles were excluded by gating as unstained particles.

Setting parameters and data analysis: Live/dead bacteria were assessed, using a EC800 Flow Cytometry Analyzer (Sony Biotechnology Inc., Champaign, IL, USA). SYTO-BC and PI were excited by a diode blue laser (488 nm). The green fluorescence emitted by bacteria with intact membrane, here designed as "live", was detected by using a 530/50 nm bandpass, and the red fluorescence emitted by bacteria with damaged membrane, here designed as "dead", was detected by using a 615/30 nm bandpass filter. Therefore, for convenience of the data analysis, double-stained bacteria corresponding to "compromised bacteria" were considered as "dead". The logarithmic mode was used to amplify the fluorescence signal. The flow rate used to analyze a sample volume of 100 µL was 10 µL/min, and a total of 40,000 cells were counted. All the samples were run in triplicate, for reproducibility of the experiment. EC800 software version 1.3.6. (Sony Biotechnology Inc., Champaign, IL, USA) was used for data analysis.

3. Results and Discussion

In this work, the synthesis of AgNPs was performed by using a well-known chemical procedure described by Lee and Meisel [21], which usually results in NPs colloidal solution that is cytotoxic in contact with mammalian cells [25,26]. The combination of AgNPs on a polymeric matrix results in the improvement of the biocompatibility, without jeopardizing the antimicrobial effect of the NPs [27,28]. That is why a nanoformulation containing chitosan solution with dispersed AgNPs to further apply on the nylon fabrics was developed. The chitosan solution was found to act as a stabilizer and dispersant of the AgNPs. As depicted in Figure 2, the AgNPs agglomerate when observed in STEM from the aqueous colloidal solution (Figure 2A), while in the presence of chitosan, they are well dispersed, keeping an average size of 24.8 $\pm$ 4.1 nm (Figure 2B). The colloidal citrate solution of synthetized AgNPs has a very low aggregation and remains stable for two months, in a refrigerator. However, due to the vacuum condition of SEM analysis (and also in the fabric during deposition), the drying effect promotes aggregation of the AgNPs. This effect is significantly lowered after the addition of chitosan that seems to act as a surfactant for the stabilization of the NPs. This is very important in terms of biomedical applications. Chitosan, besides acting as a surfactant, also possesses important antimicrobial properties. The charged amino group present in its chemical structure confers a biodegradable and biocompatible character to the mixture [27].

The AgNPs, chitosan and the mixture of both were impregnated in the nylon fabrics, through a dip-coating method, on the material with and without a previous plasma treatment. Figure 3 shows the surface of the textile material. It could be observed that the plasma-treated nylon control has a smoother surface, which is the result of a surface etching induced by the application of plasma, whose rate depends on the degree of crystallinity of the material [29]. Overall, the application of all types of antimicrobial agents (AgNPs, chitosan and the combination of both) is clearly influenced by the plasma treatment. The plasma-treated material has a higher amount of coating material on its surface, although it is not fully covered by any of the materials. The layers are more prominent on plasma-treated samples (Figure 3, right column). As expected, when the AgNPs are applied without the chitosan, they agglomerate in one part of the fiber; meanwhile, when applied together with the chitosan, a layer of this combined coating is clearly observed (Figure 3, last line).

The nylon fabric was further tested for incorporation of increased concentration of silver on the chitosan-based nanoformulations. Reflectance measurements were performed to qualitatively assess the amount of silver immobilized on the material surface. The reflectance was measured at the higher observed peak for Ag NPs (410 nm) assigned to the silver surface plasmon resonance usually located in between 405 and 425 nm [30–32]. The lower the reflectance the higher is the amount of Ag in the material surface. Thus, the decrease observed in Figure 4 on the reflectance at 410 nm indicates an increase of the NPs incorporation on the material surface mainly on the material previously treated with plasma at a concentration ratio of 60/40 or when only Ag was present (PA+CH+Ag 0/100). In fact, a significant decrease was observed when the highest concentrations of AgNPs were incorporated after plasma treatment, indicating that the improved adhesion in the presence of plasma is driven by a concentration-related mechanism (Figure 4).

Figure 5 presents the high-resolution XPS spectrum for Ag3d. As expected, the spectrum shows spin-orbit split $3d_{5/2}$ and $3d_{3/2}$ components separated by 6 eV [33]. Ag $3d_{5/2}$ and $3d_{3/2}$ binding energies are centered approximately at 368.3 $\pm$ 0.3 eV and 374.3 $\pm$ 0.3 eV, respectively. This is in good agreement with the metallic state of Ag (Ag^0) on textiles [34,35]. The rather narrow FWHM of Ag3d5/2 indicates a narrow size distribution of the nanoparticles, which is consistent with the STEM results in Figure 2 [36].

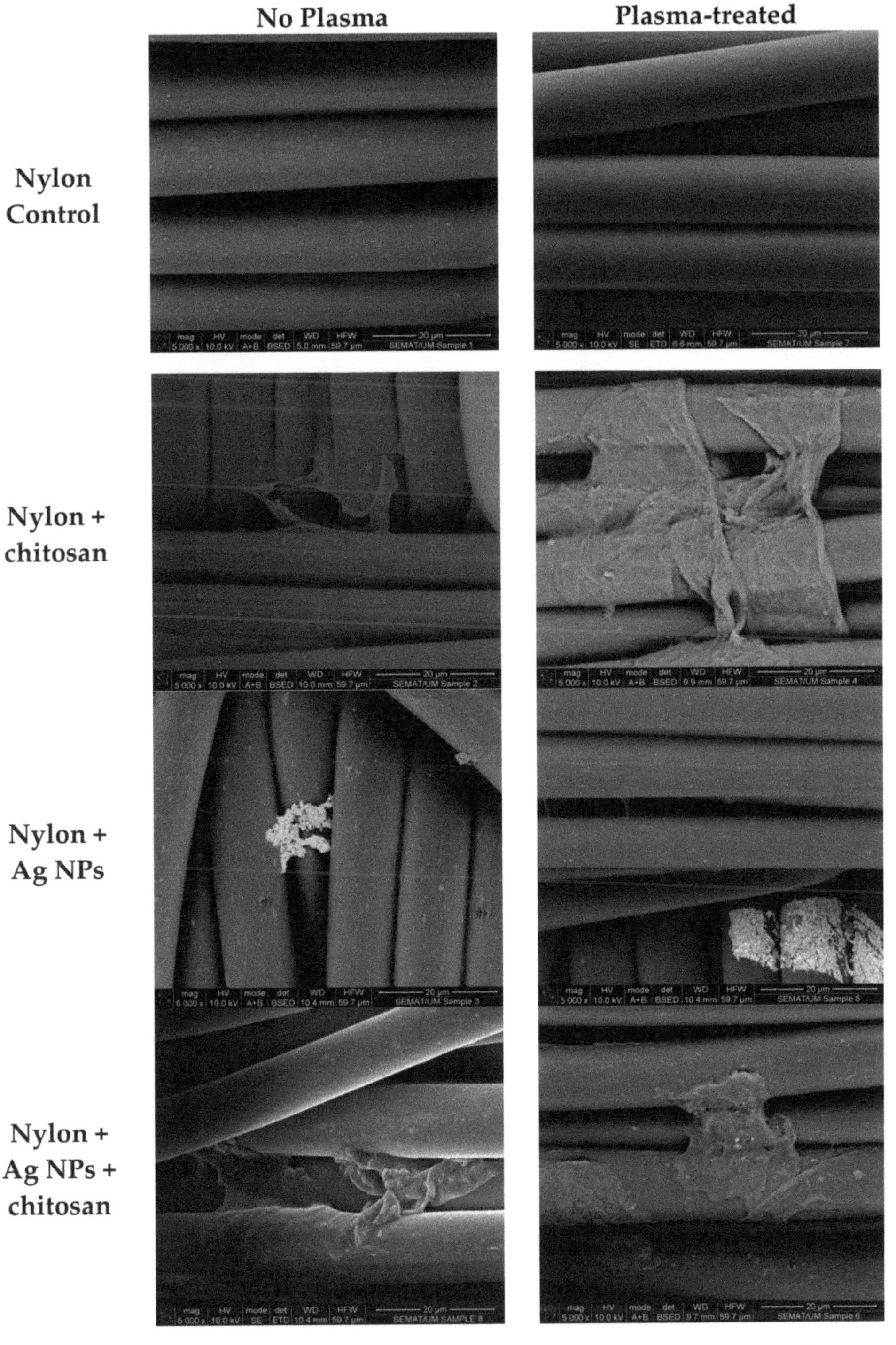

Figure 3. SEM images of the surface of nylon fabrics before (left column) and after (right column) plasma treatment. The lines depict the surface of the nylon textile prior to the functionalization (control), functionalized with chitosan (Nylon + chitosan), with Ag NPs (Nylon + Ag NPs) and with both antimicrobial agents (Nylon + Ag NPs + chitosan).

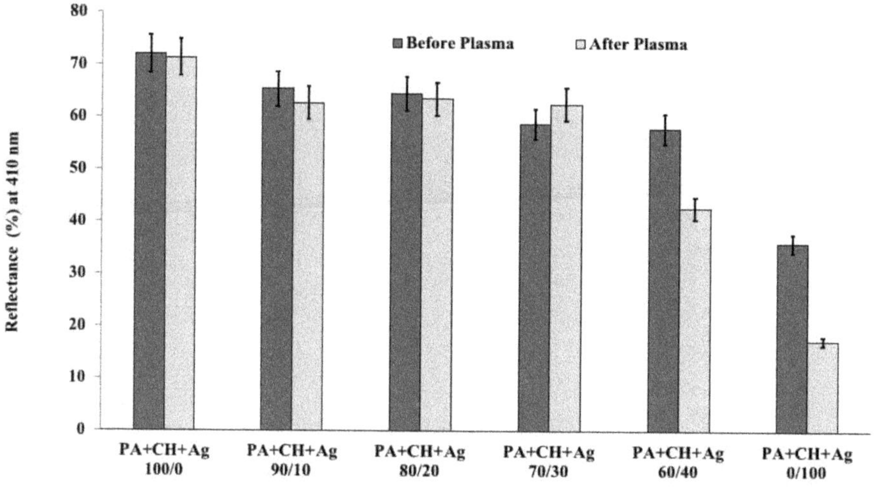

Figure 4. UV–Vis reflectance spectroscopy measurements of nylon fabric at 410 nm with increasing concentration of silver on non-treated and treated plasma nylon fabrics (PA—Nylon fabric; CH—Chitosan; Ag—Silver nanoparticles). The dilutions are represented as the percentage of chitosan and silver (CH/Ag), where the sample PA+CH+Ag 100/0 represents the control only with chitosan, and the PA+CH+Ag 0/100 represents the control only with AgNPs.

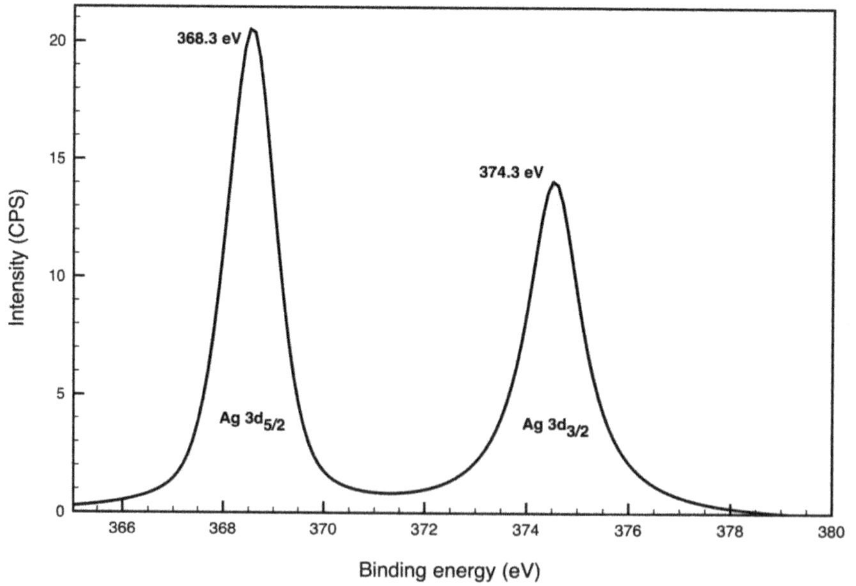

Figure 5. High-resolution XPS spectrum of the Ag3d envelope of plasma-treated PA nylon fabric with deposited AgNPs (PA + CH + Ag 0/100).

The antimicrobial property of the nylon fabric (PA) was further assessed by taking into account different combinations of chitosan (CH) and AgNPs. Overall, different antimicrobial responses were obtained for a Gram-positive bacterium (*S. aureus*) and for a Gram-negative bacterium (*P. aeruginosa*) (Figure 6). The bactericidal activity was higher for the *S. aureus* than for *P. aeruginosa* in all the combinations of chitosan and AgNPs. The percentage of *S. aureus* CFU was reduced to less than 20% when exposed to each of the combinations of chitosan and AgNPs; the fabric with percentage of chitosan and AgNPs 70/30 was the most effective. *P. aeruginosa* was less susceptible to the antimicrobial treatment; the treatment with only NPs (0/100) was ineffective. The combination of chitosan with AgNPs resulted in 60% reduction of *P. aeruginosa*, approximately. In order to evaluate the possibility of this material to be reused, the fabric was subjected to a washing treatment, after which the percentage of live bacteria increased significantly ($p < 0.001$) for all the combinations of chitosan and AgNPs. This result indicates that the chitosan and AgNPs coating is unstable, being almost completely removed from the fabrics after washing, as shown Figure 6.

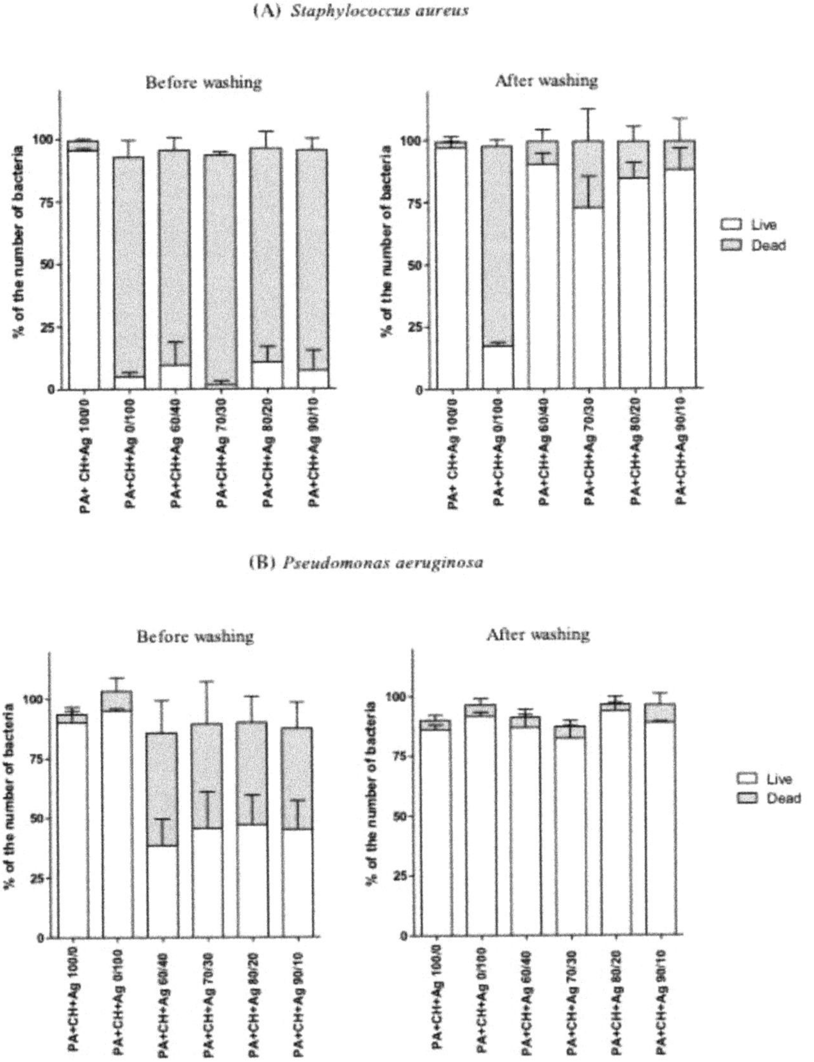

Figure 6. Antimicrobial response of *Staphylococcus aureus* (**A**) and *Pseudomonas aeruginosa* (**B**) to different combinations of chitosan (CH) and silver nanoparticles (Ag), before and after washing of the nylon fabrics (PA). The dilutions are represented as the percentage of chitosan and silver (CH/Ag), where the sample PA + CH + Ag 100/0 represents the control only with chitosan and the PA + CH + Ag 0/100 represents the control only with AgNPs.

The two bacteria tested, *S. aureus* and *P. aeruginosa*, are among the most common infection agents and representative of each group (Gram-positive and Gram-negative bacteria, respectively). As these bacteria are often found in skin and the upper respiratory tract, they will pose a serious problem in the use of facial masks, as fabric is in contact with skin for many hours. The infections caused by these two infectious agents can be minor to life-threatening, being particularly dangerous for immunocompromised patients [37–39].

P. aeruginosa is a Gram-negative bacillus characterized by a cytoplasmatic membrane with a symmetric phospholipid bilayer and asymmetric outer membrane with a phospholipid inner face and a lipopolysaccharide outer layer forming a permeability barrier [40,41]. In the presence of AgNPs, the first event is the adsorption of AgNPs into the surface of the cell membrane. This will undermine the membrane integrity, inducing the particles transport into the cytoplasm influencing several metabolic processes. It has been reported that AgNPs influence the equilibrium of the oxidation and anti-oxidation mechanisms, leading to an excess of reactive oxygen species, which can cause DNA aggregation and protein degradation, leading to the cell death [42–44]. This process, described by Liao et al., can take up to several hours; in our case, the bacterium was only in contact with the nanoparticles for 2 h, which may indicate that that was not enough contact time to obtain a higher death rate.

An important difference between Gram-negative and Gram-positive bacteria is the composition of the cell wall, which may justify the distinct effects observed on the presence of the AgNPs [45].

A Gram-positive bacterium does not have an outer membrane, similar to the Gram-negative bacterium, but has a thick layer of peptidoglycan complemented by anionic glycopolymers know as teichoic acids [46]. This structure is known to offer less resistance to the passage of some substances, like antibiotics, and it has been claimed that silver can perforate the peptidoglycan cell wall [2–4,47,48] (Figure 7).

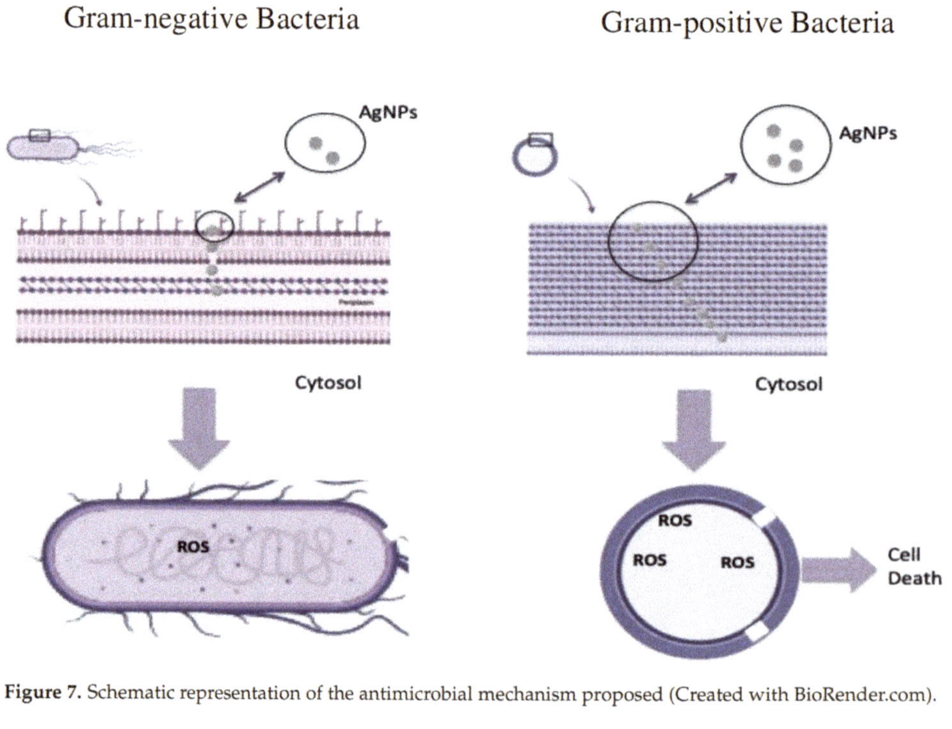

Figure 7. Schematic representation of the antimicrobial mechanism proposed (Created with BioRender.com).

4. Conclusions

This work demonstrated that, by simply using well-known antimicrobial agents such as chitosan and AgNPs, combined over a commonly used fabric such as nylon, we may obtain an antimicrobial textile suitable to be used as a face mask. The synergy between the chitosan and the AgNPs allowed for obtaining an antimicrobial coating with little amount of silver, which is a breakthrough if we considered silver toxicity. The developed synergistic coating demonstrated itself to be highly effective against the most common Gram-positive (*S. aureus*) and Gram-negative (*P. aeruginosa*) bacteria, highly decreasing their live count by approximately 80% and 60%, respectively, after 2 h. Even though the coating seems to be unstable after a washing treatment, this can be applied in the external side of facial mask fabrics, as it will prevent virus and bacteria accumulation onto the fibers. According to the literature, the presence of silver in this form may also prevent cross-contamination infections by SARS-COV-2 virus. To our knowledge, the coating of nylon fabrics intended for face masks' material with both agents has never been reported. The new-coated fabric can be successfully used for a single-use facemask.

Author Contributions: Conceptualization, J.M.S., R.F. and A.Z.; methodology, J.M.S., A.Z., M.M.F., C.M.B. and A.M.S.; formal analysis, J.M.S., C.M.B. and N.D.; investigation, J.M.S. and C.M.B.; writing—original draft preparation, J.M.S., C.M.B., A.Z. and M.M.F.; writing—review and editing, C.M.B., A.Z. and M.M.F.; supervision, J.A.T., R.F. and A.Z. All authors have read and agreed to the published version of the manuscript.

Funding: This study was supported by the Portuguese Foundation for Science and Technology (FCT), under the scope of the strategic funding of UIDB/04469/2020 unit, and BioTecNorte operation (NORTE-01-0145-FEDER-000004). funded by the European Regional Development, Fund under the scope of Norte2020—Programa Operacional Regional do Norte.

Acknowledgments: This study was supported by the Portuguese Foundation for Science and Technology (FCT), under the scope of the strategic funding of UIDB/04469/2020 unit, and BioTecNorte operation (NORTE-01-0145-FEDER-000004), funded by the European Regional Development Fund, under the scope of Norte2020—Programa Operacional Regional do Norte.

Conflicts of Interest: The authors declare no conflict of interest.

References

1. Bahadar, H.; Maqbool, F.; Niaz, K.; Abdollahi, M. Toxicity of Nanoparticles and an Overview of Current Experimental Models. *Iran. Biomed. J.* **2016**, *20*, 1–11. [CrossRef] [PubMed]
2. Cheng, L.; Zhang, K.; Zhang, N.; Melo, M.A.S.; Weir, M.D.; Zhou, X.D.; Bai, Y.X.; Reynolds, M.A.; Xu, H.H.K. Developing a New Generation of Antimicrobial and Bioactive Dental Resins. *J. Dent. Res.* **2017**, *96*, 855–863. [CrossRef] [PubMed]
3. Cinteza, L.O.; Scomoroscenco, C.; Voicu, S.N.; Nistor, C.L.; Nitu, S.G.; Trica, B.; Jecu, M.-L.; Petcu, C. Chitosan-Stabilized Ag Nanoparticles with Superior Biocompatibility and Their Synergistic Antibacterial Effect in Mixtures with Essential Oils. *Nanomaterials* **2018**, *8*, 826. [CrossRef] [PubMed]
4. Regiel-Futyra, A.; Kus-Liśkiewicz, M.; Sebastian, V.; Irusta, S.; Arruebo, M.; Kyzioł, A.; Stochel, G. Development of noncytotoxic silver–chitosan nanocomposites for efficient control of biofilm forming microbes. *RSC Adv.* **2017**, *7*, 52398–52413. [CrossRef]
5. Elechiguerra, J.L.; Burt, J.L.; Morones, J.R.; Camacho-Bragado, A.; Gao, X.; Lara, H.H.; Yacaman, M.J. Interaction of silver nanoparticles with HIV-1. *J. Nanobiotechnol.* **2005**, *3*, 6. [CrossRef] [PubMed]
6. Sun, R.W.; Chen, R.; Chung, N.P.; Ho, C.M.; Lin, C.L.; Che, C.M. Silver nanoparticles fabricated in Hepes buffer exhibit cytoprotective activities toward HIV-1 infected cells. *Chem. Commun.* **2005**, 5059–5061. [CrossRef]
7. Lu, L.; Sun, R.W.; Chen, R.; Hui, C.K.; Ho, C.M.; Luk, J.M.; Lau, G.K.; Che, C.M. Silver nanoparticles inhibit hepatitis B virus replication. *Antivir. Ther.* **2008**, *13*, 253–262.
8. Baram-Pinto, D.; Shukla, S.; Perkas, N.; Gedanken, A.; Sarid, R. Inhibition of Herpes Simplex Virus Type 1 Infection by Silver Nanoparticles Capped with Mercaptoethane Sulfonate. *Bioconjug. Chem.* **2009**, *20*, 1497–1502. [CrossRef]
9. Sun, L.; Singh, A.; Vig, K.; Pillai, S.; Singh, S. Silver Nanoparticles Inhibit Replication of Respiratory Syncytial Virus. *J. Biomed. Nanotechnol.* **2008**, *4*, 149–158.
10. Rogers, J.V.; Parkinson, C.V.; Choi, Y.W.; Speshock, J.L.; Hussain, S.M. A Preliminary Assessment of Silver Nanoparticle Inhibition of Monkeypox Virus Plaque Formation. *Nanoscale Res. Lett.* **2008**, *3*, 129–133. [CrossRef]
11. Xiang, D.X.; Chen, Q.; Pang, L.; Zheng, C.L. Inhibitory effects of silver nanoparticles on H1N1 influenza A virus in vitro. *J. Virol. Methods* **2011**, *178*, 137–142. [CrossRef] [PubMed]

12. Zachar, O. Formulations for COVID-19 Treatment via Silver Nanoparticles Inhalation Delivery at Home and Hospital. *ScienceOpen Preprints* **2020**, 1. [CrossRef]
13. Ivanova, A.A.; Surmenev, R.A.; Surmeneva, M.A.; Mukhametkaliyev, T.; Loza, K.; Prymak, O.; Epple, M. Hybrid biocomposite with a tunable antibacterial activity and bioactivity based on RF magnetron sputter deposited coating and silver nanoparticles. *Appl. Surf. Sci.* **2015**, *329*, 212–218. [CrossRef]
14. Alissawi, N.; Peter, T.; Strunskus, T.; Ebbert, C.; Grundmeier, G.; Faupel, F. Plasma-polymerized HMDSO coatings to adjust the silver ion release properties of Ag/polymer nanocomposites. *J. Nanopart. Res.* **2013**, *15*. [CrossRef]
15. Pappas, D. Status and potential of atmospheric plasma processing of materials. *J. Vac. Sci. Technol. A* **2011**, *29*, 020801. [CrossRef]
16. Jia, C.X.; Chen, P.; Liu, W.; Li, B.; Wang, Q.A. Surface treatment of aramid fiber by air dielectric barrier discharge plasma at atmospheric pressure. *Applied Surface Science* **2011**, *257*, 4165–4170. [CrossRef]
17. Radić, N.; Obradović, B.M.; Kostić, M.; Dojčinović, B.; Kuraica, M.M.; Černák, M. Deposition of silver ions onto DBD and DCSBD plasma treated nonwoven polypropylene. *Surf. Coat. Technol.* **2012**, *206*, 5006–5011. [CrossRef]
18. Zille, A.; Oliveira, F.R.; Souto, A.P. Plasma Treatment in Textile Industry. *Plasma Process. Polym.* **2015**, *12*, 98–131. [CrossRef]
19. Hosne Asif, A.K.M.A.; Hasan, M.Z. Application of Nanotechnology in Modern Textiles: A Review. *Int. J. Curr. Eng. Technol.* **2018**, *8*. [CrossRef]
20. Mei, L.; Xu, Z.; Shi, Y.; Lin, C.; Jiao, S.; Zhang, L.; Li, P. Multivalent and synergistic chitosan oligosaccharide-Ag nanocomposites for therapy of bacterial infection. *Sci. Rep.* **2020**, *10*, 1–9. [CrossRef]
21. Lee, P.C.; Meisel, D. Adsorption and surface-enhanced Raman of dyes on silver and gold sols. *J. Phys. Chem.* **1982**, *86*, 3391–3395. [CrossRef]
22. Oliveira, F.; Souto, A.; Carneiro, N.; Oliveira, H. Surface Modification on Polyamide 6.6 with Double Barrier Discharge (DBD) Plasma to Optimise Dyeing Process by Direct Dyes. *Mater. Sci. Forum* **2010**, *636–637*, 846–852. [CrossRef]
23. Zille, A.; Fernandes, M.M.; Francesko, A.; Tzanov, T.; Fernandes, M.; Oliveira, F.R.; Almeida, L.; Amorim, T.; Carneiro, N.; Esteves, M.F.; et al. Size and Aging Effects on Antimicrobial Efficiency of Silver Nanoparticles Coated on Polyamide Fabrics Activated by Atmospheric DBD Plasma. *ACS Appl. Mater. Interfaces* **2015**, *7*, 13731–13744. [CrossRef] [PubMed]
24. Oliveira, F.R.; Zille, A.; Souto, A.P. Dyeing mechanism and optimization of polyamide 6,6 functionalized with double barrier discharge (DBD) plasma in air. *Appl. Surf. Sci.* **2014**, *293*, 177–186. [CrossRef]
25. Wang, L.-S.; Wang, C.-Y.; Yang, C.-H.; Hsieh, C.-L.; Chen, S.-Y.; Shen, C.-Y.; Wang, J.-J.; Huang, K.-S. Synthesis and anti-fungal effect of silver nanoparticles-chitosan composite particles. *Int. J. Nanomed.* **2015**, *10*, 2685–2696. [CrossRef]
26. Sambale, F.; Wagner, S.; Stahl, F.; Khaydarov, R.R.; Scheper, T.; Bahnemann, D. Investigations of the Toxic Effect of Silver Nanoparticles on Mammalian Cell Lines. *J. Nanomater.* **2015**, *2015*, 136765. [CrossRef]
27. Regiel, A.; Irusta, S.; Kyzioł, A.; Arruebo, M.; Santamaria, J. Preparation and characterization of chitosan-silver nanocomposite films and their antibacterial activity against Staphylococcus aureus. *Nanotechnology* **2013**, *24*, 015101. [CrossRef]
28. Pakseresht, S.; Alogaili, A.W.M.; Akbulut, H.; Placha, D.; Pazdziora, E.; Klushina, D.; Konvičková, Z.; Kratošová, G.; Holešová, S.; Martynková, G.S. Silver/Chitosan Antimicrobial Nanocomposites Coating for Medical Devices: Comparison of Nanofiller Effect Prepared via Chemical Reduction and Biosynthesis. *J. Nanosci. Nanotechnol.* **2019**, *19*, 2938–2942. [CrossRef]
29. Neděla, O.; Slepička, P.; Švorčík, V. Surface Modification of Polymer Substrates for Biomedical Applications. *Materials* **2017**, *10*, 1115. [CrossRef]
30. Khodashenas, B.; Ghorbani, H.R. Synthesis of silver nanoparticles with different shapes. *Arab. J. Chem.* **2019**, *12*, 1823–1838. [CrossRef]
31. Carré-Rangel, L.; Alonso-Nuñez, G.; Espinoza-Gómez, H.; Flores-López, L.Z. Green Synthesis of Silver Nanoparticles: Effect of Dextran Molecular Weight Used as Stabilizing-Reducing Agent. *J. Nanosci. Nanotechnol.* **2015**, *15*, 9849–9855. [CrossRef] [PubMed]
32. Bankura, K.P.; Maity, D.; Mollick, M.M.R.; Mondal, D.; Bhowmick, B.; Bain, M.K.; Chakraborty, A.; Sarkar, J.; Acharya, K.; Chattopadhyay, D. Synthesis, characterization and antimicrobial activity of dextran stabilized silver nanoparticles in aqueous medium. *Carbohydr. Polym.* **2012**, *89*, 1159–1165. [CrossRef] [PubMed]
33. Do, J.Y.; Chava, R.K.; Kim, Y.I.; Cho, D.W.; Kang, M. Fabrication of Ag based ternary nanocomposite system for visible-light photocatalytic hydrogen evolution reaction. *Appl. Surf. Sci.* **2019**, *494*, 886–894. [CrossRef]
34. McNally, M.J.; Galinis, G.; Youle, O.; Petr, M.; Prucek, R.; Machala, L.; von Haeften, K. Silver nanoparticles by atomic vapour deposition on an alcohol micro-jet. *Nanoscale Adv.* **2019**, *1*, 4041–4051. [CrossRef]
35. Zhu, H.; Chen, D.; Li, N.; Xu, Q.; Li, H.; He, J.; Lu, J. Cyclodextrin-functionalized Ag/AgCl foam with enhanced photocatalytic performance for water purification. *J. Colloid Interface Sci.* **2018**, *531*, 11–17. [CrossRef]
36. Algarra, M.; Campos, B.B.; Radotić, K.; Mutavdžić, D.; Bandosz, T.; Jiménez-Jiménez, J.; Rodriguez-Castellón, E.; Esteves da Silva, J.C.G. Luminescent carbon nanoparticles: Effects of chemical functionalization, and evaluation of Ag+ sensing properties. *J. Mater. Chem. A* **2014**, *2*. [CrossRef]
37. Klastersky, J. Nosocomial infections due to Gram-negative bacilli in compromised hosts: Considerations for prevention and therapy. *Rev. Infect. Dis.* **1985**, *7* (Suppl. 4), S552–S558. [CrossRef]
38. Wu, D.; Chan, W.; Metelitsa, A.; Fiorillo, L.; Lin, A. Pseudomonas skin infection: Clinical features, epidemiology, and management. *Am. J. Clin. Dermatol.* **2011**, *12*, 157–169. [CrossRef]

39. Dayan, G.H.; Mohamed, N.; Scully, I.L.; Cooper, D.; Begier, E.; Eiden, J.; Jansen, K.U.; Gurtman, A.; Anderson, A.S. Staphylococcus aureus: The current state of disease, pathophysiology and strategies for prevention. *Expert Rev. Vaccines* **2016**, *15*, 1373–1392. [CrossRef]
40. Chevalier, S.; Bouffartigues, E.; Bodilis, J.; Maillot, O.; Lesouhaitier, O.; Feuilloley, M.G.J.; Orange, N.; Dufour, A.; Cornelis, P. Structure, function and regulation of Pseudomonas aeruginosa porins. *FEMS Microbiol. Rev.* **2017**, *41*, 698–722. [CrossRef]
41. Liao, S.; Zhang, Y.; Pan, X.; Zhu, F.; Jiang, C.; Liu, Q.; Cheng, Z.; Dai, G.; Wu, G.; Wang, L.; et al. Antibacterial activity and mechanism of silver nanoparticles against multidrug-resistant Pseudomonas aeruginosa. *Int. J. Nanomed.* **2019**, *14*, 1469–1487. [CrossRef] [PubMed]
42. Abdal Dayem, A.; Hossain, M.K.; Lee, S.B.; Kim, K.; Saha, S.K.; Yang, G.-M.; Choi, H.Y.; Cho, S.-G. The Role of Reactive Oxygen Species (ROS) in the Biological Activities of Metallic Nanoparticles. *Int. J. Mol. Sci.* **2017**, *18*, 120. [CrossRef] [PubMed]
43. Mao, B.-H.; Chen, Z.-Y.; Wang, Y.-J.; Yan, S.-J. Silver nanoparticles have lethal and sublethal adverse effects on development and longevity by inducing ROS-mediated stress responses. *Sci. Rep.* **2018**, *8*, 2445. [CrossRef] [PubMed]
44. Cameron, S.J.; Hosseinian, F.; Willmore, W.G. A Current Overview of the Biological and Cellular Effects of Nanosilver. *Int. J. Mol. Sci.* **2018**, *19*, 2030. [CrossRef] [PubMed]
45. Sobhanifar, S.; Worrall, L.J.; King, D.T.; Wasney, G.A.; Baumann, L.; Gale, R.T.; Nosella, M.; Brown, E.D.; Withers, S.G.; Strynadka, N.C.J. Structure and Mechanism of Staphylococcus aureus TarS, the Wall Teichoic Acid β-glycosyltransferase Involved in Methicillin Resistance. *PLoS Pathog.* **2016**, *12*, e1006067. [CrossRef] [PubMed]
46. Sobhanifar, S.; Worrall, L.J.; Gruninger, R.J.; Wasney, G.A.; Blaukopf, M.; Baumann, L.; Lameignere, E.; Solomonson, M.; Brown, E.D.; Withers, S.G.; et al. Structure and mechanism of Staphylococcus aureus TarM, the wall teichoic acid α-glycosyltransferase. *Proc. Natl. Acad. Sci. USA* **2015**, *112*, E576–E585. [CrossRef]
47. Sim, W.; Barnard, R.T.; Blaskovich, M.A.T.; Ziora, Z.M. Antimicrobial Silver in Medicinal and Consumer Applications: A Patent Review of the Past Decade (2007–2017). *Antibiotics* **2018**, *7*, 93. [CrossRef]
48. Lambert, P.A. Cellular impermeability and uptake of biocides and antibiotics in Gram-positive bacteria and mycobacteria. *J. Appl. Microbiol.* **2002**, *92*, 46s–54s. [CrossRef]

fibers

Article

Synthesis and Characterization of SiO_2@CNTs Microparticles: Evaluation of Microwave-Induced Heat Production

Panagiotis Kainourgios, Ioannis A. Kartsonakis * and Costas A. Charitidis *

Research Unit of Advanced, Composite, Nano-Materials and Nanotechnology, School of Chemical Engineering, National Technical University of Athens, 9 Heroon Polytechniou Street, Zographos, GR-15773 Athens, Greece; pkainourgios@chemeng.ntua.gr

* Correspondence: ikartso@chemeng.ntua.gr (I.A.K.); charitidis@chemeng.ntua.gr (C.A.C.); Tel.: +30-210-772-4046 (C.A.C.)

Abstract: This study was focused on the growth of multi-walled carbon nanotubes (MWCNTs) on iron chloride-functionalized silica microspheres. In addition, the microwave absorption potential and the subsequent heat production of the resulting structures were monitored by means of infrared thermometry and compared with pristine commercially available MWCNTs. The functionalized silica microparticle substrates produced MWCNTs without any amorphous carbon but with increased structural defects, whereas their heat production performance as microwave absorbents was comparable to that of the pristine MWCNTs. Two-minute microwave irradiation of the SiO_2@CNTs structures resulted in an increase in the material's temperature from ambient temperature up to 173 °C. This research puts forward a new idea of charge modulation of MWCNTs and sheds light on an investigation for the development of bifunctional materials with improved properties with respect to efficient microwave absorbance.

Keywords: microwave irradiation; carbon nanotubes; chemical vapor deposition; nanostructured powders; Raman

check for updates

Citation: Kainourgios, P.; Kartsonakis, I.A.; Charitidis, C.A. Synthesis and Characterization of SiO_2@CNTs Microparticles: Evaluation of Microwave-Induced Heat Production. *Fibers* **2021**, *9*, 81. https://doi.org/10.3390/fib9120081

Academic Editor: Ramiro Rafael Ruiz Rosas

Received: 4 November 2021
Accepted: 1 December 2021
Published: 3 December 2021

1. Introduction

Currently, reduction, recycling and reuse are crucial parameters in order to tackle waste problems and to move from linear economic processes and systems towards a more circular economy [1]. Therefore, material development methods progressively lean towards green synthesis procedures, aiming at recycling strategies as well as the reusability of raw materials. Furthermore, in the industries related to composite and coating manufacture, there is a growing demand for reclaimable materials, which opposes a serious challenge due to the permanent nature of the bonding between the matrix and reinforcing material [1]. For many years, the term "debonding" has been addressed as a form of material failure, where loss of adhesion occurs between the matrix and filler [2]. On the other hand, recent studies propose alternatives for the fabrication of coatings where debonding-on-demand is being described as a mechanism to separate the two phases of a composite material or a coated surface via the application of an external stimulus. From the chemical perspective, such strategies include the synthesis of crosslinked polymers utilizing photo-initiators for triggering mechanisms [3–5]. In addition, the use of nanoparticles (NPs) could potentially address this challenge due to their unique intrinsic properties [6–8].

Notwithstanding, CNTs have attracted significant attention as filler materials because of their unique physical properties [9]. One of the most important properties of CNTs is their response to microwave irradiation by producing heat. Imholt et al. previously described the emission of light as well as the production of intense heat when single-walled carbon nanotubes (SWCNTs) were exposed to microwave irradiation [10]. For these reasons, several vibration analyses have been performed related to the number of walls of the nanotubes. In the work of Strozzi et al., linear vibrations of triple-walled carbon nanotubes were investigated in order to determine the effect of the geometry and boundary conditions on the natural frequencies of the nanotubes [11]. In the study of He et al., it was found that the van der Waals interaction influences the radial vibration of small-radius multi-walled carbon nanotubes (MWCNTs) [12]. Moreover, the mechanical behavior of MWNTs as single entities, as well as their effectiveness as load-bearing entities in nanocomposite materials, was investigated by Pantano et al., focusing on the use of bent MWNTs of different diameters and numbers of walls [13]. Although the exact mechanism of microwave absorbance is still under vigorous research, its interaction is also being studied in terms of microwave-assisted synthesis, functionalization or purification, whereas in all cases examined, the production of heat is a common aspect [14].

Purification of MWCNTs includes several steps and can be performed via a variety of procedures that emphasize the removal of the catalytic particles which are important for the growth of CNTs. These particles are transition metal nanoparticles, in most cases iron oxide nanoparticles. In the work of Wadhawan et al., the microwave absorption of purified and unpurified SWCNTs was examined. The outcome of their study revealed an increase in heating in the unpurified sample compared to the purified one. They also found that the unpurified SWCNTs also emitted light, a result that clearly denoted the evidence of heat production after microwave irradiation, as well as the hypothesis that iron oxide nanoparticles contribute to microwave absorption [15]. The exceptional properties of CNTs are attributed to their defect-free graphene layers, including microwave absorption; the defect-free surface, however, renders the material chemically inert and therefore opposes a challenge towards practical applications such as homogenous dispersions of CNTs in thermoplastic matrices [16].

On the other hand, silica nanoparticles produced by the Stöber process, can be surface functionalized in order to be homogenously dispersed in a variety of media [17]. In the work of Wang et al., a core–shell three-dimensional hierarchical structure based on Fe_3O_4 was fabricated in a multi-component glass system that was able to exhibit effective microwave absorption [18]. In addition, Green et al. produced doped silica nanoparticles, which, according to their findings, exhibited excellent microwave absorption performance [19]. From the aforementioned research outcomes, it is evident that both silica and iron oxide have potential microwave absorption properties.

In this study, silica nanoparticles were fabricated and surface functionalized with iron chloride to provide substrates/catalytic particles for the growth of carbon nanotubes. The CNTs' growth was studied morphologically (and qualitatively) by scanning electron microscopy (SEM), quantitatively by means of thermogravimetric analysis (TGA) and qualitatively utilizing Raman spectroscopy to identify the I_d/I_g ratio. The as-produced nanomaterials were subsequently microwave irradiated, and their heat production was estimated by incorporating infrared thermometry. The added value of this work lies in the fact that the individual properties of each component, namely, surface functionalization and microwave absorption for silica nanoparticles and MWCNTs, respectively, render this hybrid material a potential debonding agent in the composite and coating industries. The resulting hybrid structures present both microsilica properties (high surface area, surface functionalization/chemical affinity with polymer matrices) and heat production via microwave absorption due to the MWCNT counterpart and therefore can be considered a novel material in the composite and coating industries.

2. Materials and Methods

2.1. Materials

Absolute ethanol 99.5% and ammonium hydroxide were purchased from Acros Organics, whereas tetraethyl orthosilicate (TEOS) and iron chloride (II) tetrahydrate 99% were purchased from Sigma Aldrich (Saint Louis, MI, USA) and used without any further purification. For the chemical vapor deposition (CVD) experiment, acetylene was used as the carbon source, whereas argon and hydrogen were used as inert and reducing agents, respectively. Finally, apart from the laboratory synthesis of MWCNTs, pristine commercial MWCNTs were also purchased from Hongwu International Group Ltd. (Guangzhou, China).

2.2. Synthesis of SiO_2@CNTs

The synthesis of MWCNTs on the surface of colloidal silica, SiO_2@CNTs, was accomplished by a three-step process: (i) formation of colloidal silica by means of sol-gel (Stöber method) [17], (ii) functionalization of the hydroxyl groups of colloidal silica with iron chloride and iii) heat treatment of the functionalized SiO_2 at 700 °C under an oxygen atmosphere to calcinate the sample and form iron oxide nanoparticles, with subsequent reduction under H_2 flow in the presence of acetylene as a carbon source to initiate the growth of carbon nanotubes.

Briefly, 100 mL of ethanol, 11 mL of H_2O, 6 mL of TEOS and 8 mL of ammonia as a catalyst were placed in a 250 mL round-bottom flask. The mixture was left under vigorous stirring for 5 h to produce silica microspheres with average diameters of 350 nm. Subsequently, centrifugation and rinsing with deionized water followed. The as-received white powder was left to dry in a desiccator.

Functionalization of the hydroxyl groups of colloidal silica was conducted with iron chloride aqueous solution by means of electrostatic interaction. The mechanism of the electrostatic interaction is realized by the interaction of the negatively charged hydroxyl groups of silica with the positively charged Fe^+ ions. This technique was demonstrated in our previous work utilizing negatively charged polymethacrylic acid nanoparticles and silver ions [20].

The silica nanospheres were dispersed in an aqueous solution of 10 mM iron chloride tetrahydrate 99% for 2 h in order for Fe^+ ions to electrostatically attach to the hydroxyl groups of silica. Centrifugation and rinsing followed, and the as-received orange powder was dried in a desiccator.

After the functionalized nanospheres were dried, they were grinded into a fine powder and placed on top of a silicon wafer. Then, the wafer was placed inside a horizontal CVD furnace that was set to 700 °C under an oxygen atmosphere to calcinate the sample and form iron oxide nanoparticles. Subsequently, a reduction process followed under H_2 flow (220 sccm) for 10 min, and then under acetylene flow (300 sccm) for 10 min as a carbon source inserted into the system to initiate the growth of MWCNTs on the colloidal silica surface. Further purification (removal of the iron oxide nanoparticles by chemical oxidation) of the as-produced MWCNTs was not realized in order to avoid introducing further defects in the MWCNTs.

2.3. Characterization

The materials' surface morphology was estimated via scanning electron microscopy (SEM) using a PHILIPS Quanta Inspect (FEI Company, Hillsboro, OR, USA) microscope with a W (tungsten) filament 25 KV equipped with a EDAX GENESIS (Ametex Process & Analytical Instruments, Pittsburgh, PA, USA). Micro-Raman measurements were performed using a Renishaw inVia spectrometer working in a backscattering configuration and equipped with a near-infrared diode laser emitting at 530 nm. The spectra were recorded by focusing the laser beam on the sample surface and adjusting the light power so that 1 mW was provided for a spot of about 1 μm diameter. Thermogravimetric analysis (TGA) was performed on a TGA Netch STA 449 Jupiter. Samples were heated from ambient temperature to 900 °C with a heating rate of 10 °C/min under an O_2 atmosphere. A Malvern Zetasizer nano zs apparatus was utilized for measuring the size, polydispersity and surface charge of the silica microsphere by dynamic light scattering (DLS). In addition, a Cary 630 FT-IR Spectrometer was utilized to perform the FT-IR measurements. Each sample was grinded into a powder and mixed with KBr to form pellets for the measurements in the range of 400 to 4500 cm^{-1}.

Microwave absorption of materials is a complex scientific field that requires a well-established profile of the electromagnetic properties of the tested material in order to be assessed with precision. This work was focused on the synthesis and characterization of a novel material with potential microwave absorption properties; the method applied for the microwave absorption is straightforward, and the presented results serve as an indication and not as absolute values. The microwave absorption was measured utilizing a microwave instrument (magnetron 700 W operating at 2.45 GHz), whereas the heat produced was assessed utilizing an FLIR C3-X thermal camera. For the microwave absorption assessment, 4 samples were tested: an empty vial as a control, plain silica microspheres, pristine commercial MWCNTs, with a mass corresponding to a 20% mass percentage of the hybrid structures, and the final hybrid structure, which, from this point on, will be referred to as SiO_2@CNTs.

3. Results and Discussion

3.1. Morphological Characterization

The morphological evaluation of SiO_2@CNTs was accomplished via SEM, whereas DLS was used to correlate the size results with SEM analysis; the results are exhibited in Figures 1 and 2, respectively. Figure 1a depicts the pristine silica microspheres with an average diameter at 350 nm; these results are in agreement with the size measurements obtained from DLS, Figure 2a, which reveal an average hydrodynamic radius of 400 nm. The 50 nm difference can be attributed to the strong negative charge of silica microspheres (−40 mV), as seen in the zeta potential measurement in Figure 2b. Furthermore, the formation of iron oxide nanoparticles on the surface of silica microspheres prior to the growth of the MWCNTs is demonstrated in Figure 1b,c. It can be seen in Figure 1c that iron oxide nanoparticles with a diameter range from 15 to 21 nm were grown on the silica microspheres, after functionalizing the latter with iron chloride and subsequent thermal oxidation at 700 °C. However, the aforementioned observation is not evident for all the silica microspheres since the electrostatic attachment is of a dynamic nature, affected by the surface charge of the silica particles, the concentration of iron chloride and the experimental conditions such as the temperature and stirring time. Energy-dispersive X-ray analysis (EDS) was performed on the silica microspheres prior to the growth of MWCNTs; the results are illustrated in Figure 3. It can be observed that the elemental analysis affirms the presence of elemental iron at 6.3 and 0.7 keV. However, it should be mentioned that the presence of carbon and gold is attributed to the SEM sample preparation: carbon due to the use of carbon tape for sample adhesion, and gold due to the gilding process in order for the sample to become more conductive. Therefore, carbon and gold were excluded from the elemental quantification and the corresponding calculation of the atomic percentages that are illustrated in Figure 3.

Figure 1d–f exhibit the surface iron oxide-functionalized silica microspheres after the growth of MWCNTs. Starting from smaller magnifications, Figure 1d, moving to larger ones, Figure 1f, the growth of MWCNTs with a random orientation and an average thickness of 50 nm can be observed. As in the case of iron oxide nanoparticles, some silica microspheres do not present any MWCNT growth, Figure 1e,f. This outcome can be potentially attributed to partial functionalization of the initial silica microspheres due to the dynamic nature of the electrostatic bonding of Fe ions to the negatively charged microspheres [21,22]. Nevertheless, both the iron oxide nanoparticles and the as-produced MWCNTs exhibit an extremely narrow nanoparticle size and MWCNT thickness distribution, thus further supporting the effect of the catalytic nanoparticle size in the growth of MWCNTs [21,23].

Figure 1. SEM images of (**a**) silica microspheres; (**b**,**c**) silica microspheres with iron oxide nanoparticles at different magnifications; and (**d**–**f**) SiO_2@CNTs at different magnifications.

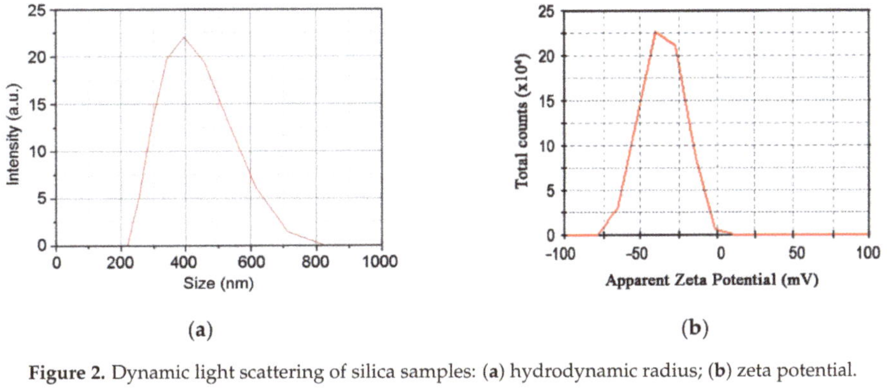

Figure 2. Dynamic light scattering of silica samples: (**a**) hydrodynamic radius; (**b**) zeta potential.

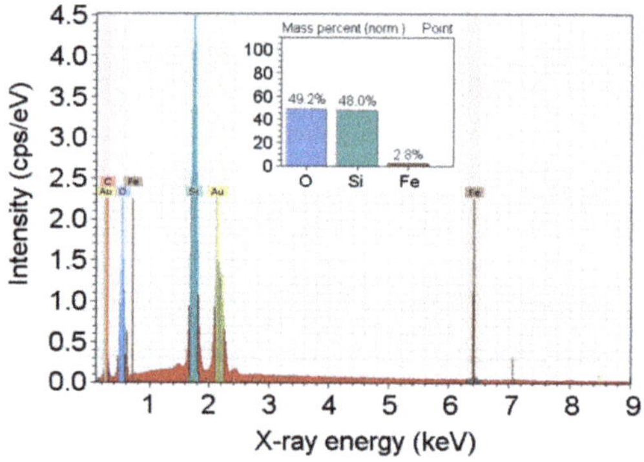

Figure 3. EDS analysis of the Fe^{+}-functionalized silica microspheres.

3.2. Structural and Thermal Analyses

Raman and TGA analyses were performed for the quantitative determination and qualitative identification, respectively, of the structural defects of MWCNTs on the SiO_2@CNTs hybrid structures. Figure 4 presents the results of the TGA analysis. It is shown that at approximately 100 °C, a minor mass increase can be observed (0.8 wt.%), which is attributed to the initial oxidation. According to Mansfield et al., amorphous carbon thermal degradation occurs between 200 and 400 °C. Therefore, taking into account that no mass loss can be seen in this temperature range, it can be assumed that the SiO_2@CNTs sample has no amorphous carbon [24,25]. The thermal degradation of the sample initiates at 463 °C and ends at 670 °C, having lost 20 wt.% of its total mass. From there on, the mass remains constant up to 900 °C, where the measurement is finished. This mass loss is attributed to the thermal oxidation of MWCNTs, which occurs above 450 °C in air [26]. According to the literature, above the aforementioned temperature under an oxygen atmosphere, the graphite layers of MWCNTs become unstable and begin to degrade; this outcome has been observed during the heating of MWCNTs at various heating rates [27]. Evidently, the TGA measurements revealed a 20 wt.% MWCNTs content with minor to no amorphous carbon content.

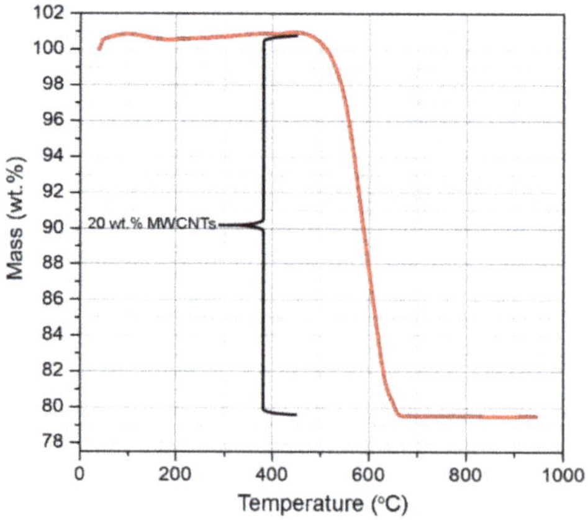

Figure 4. TGA curve of the thermal decomposition of SiO_2@CNTs.

Although the SiO_2@CNTs were free of amorphous carbon, Raman analysis (Figure 5a) revealed significant structural defects in the carbon structures judging by the calculation of the I_d/I_g ratio (I_d/I_g = 2.0). In addition to the D and G bands, the 2D (or D′) band is also present, along with the D + G band, thus further justifying the presence of MWCNTs in the sample. On the other hand, the Raman spectrum of the pristine commercial MWCNTs, Figure 5b, presents the same bands; however, the D band at approximately 1350 cm^{-1}, which is correlated with defects in the CNTs' graphitic structures, is significantly lower in area compared to the SiO_2@CNTs spectrum, Figure 5a, as well as the I_d/I_g ratio (0.6), whereas the 2D band is notably stronger. It is evident, from the Raman analysis, that the commercial MWCNTs sample exhibits a more refined structure in terms of defects in the graphitic layers.

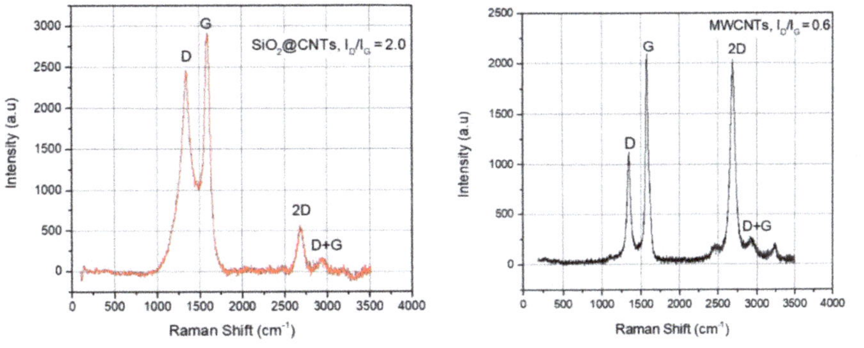

(**a**) (**b**)

Figure 5. Raman shift of (**a**) SiO_2@CNTs and (**b**) pristine commercial MWCNTs.

To further support the presence of MWCNTs on the as-produced silica surface, XRD measurements took place in order to identify any crystal structures present on the sample as evidence of MWCNTs; the results are presented in Figure 6. The XRD spectrum exhibits one broad and one sharp peak located at approximately 25 degrees and 43 degrees, respectively. According to the literature [28], amorphous nano-silica exhibits a broad peak with a center at approximately 25 degrees, which agrees with our findings since in the TGA measurements, it was shown that the sample consists primarily of silica (80%). Nevertheless, MWCNTs exhibit two main peaks, the first one located at 23 degrees and the second one at 43 degrees, responsible for the diffraction of the crystal graphitic planes (002) and (100), respectively [29,30]. Although the first peak is most likely overlapped by the strong diffraction of the amorphous silica, the second peak at 43 degrees is a sharp peak with a strong intensity, clearly denoting evidence of crystal structures in the SiO_2@CNTs sample.

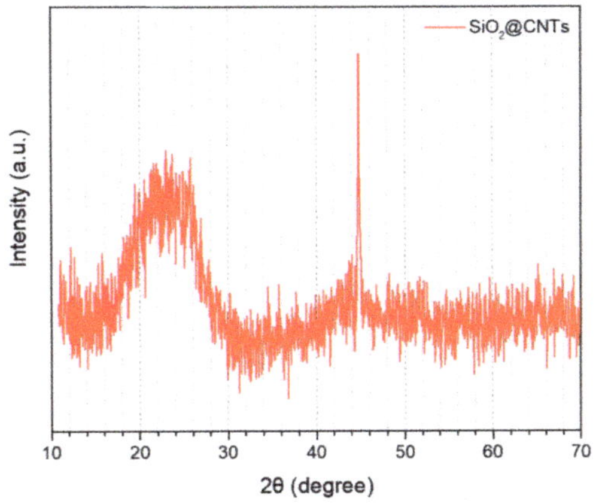

Figure 6. XRD spectrum of SiO_2@CNTs.

According to the literature, the catalytic activity for the growth of MWCNTs is provided by transition metal particles—in our case, elemental iron [31,32]. For this reason, after the functionalization of the hydroxyl groups of colloidal silica with iron chloride aqueous solution by means of electrostatic interaction, the produced iron oxide nanoparticles undergo reduction to elemental iron during the reduction process with H_2 at 700 °C. FT-IR analysis was performed on the pristine silica microspheres and on the silica microspheres after the functionalization with iron chloride and their calcination at 700 °C (Fe^+—functionalized silica microspheres). The results are illustrated in Figure 7. Figure 7a exhibits the FT-IR spectrum of the pristine silica microspheres; the absorption band at 810 cm^{-1} is attributed to Si–O–Si stretching. The absorption bands at 1070 cm^{-1} and 950 cm^{-1} are assigned to the siloxane vibration, whereas the bands at 3200–3500 cm^{-1} and at 1625 cm^{-1} are attributed to the O–H stretching band of either the hydroxyl groups of the silica microspheres or the water molecules present in the sample [33]. Figure 7b depicts the FT-IR absorption spectrum of the silica microspheres after the growth of iron oxide nanoparticles. The spectrum exhibits all the previously observed bands of the silica structure with the addition of one more peak located at 570 cm^{-1}, which, according to the literature, is an indication of the vibration of Si–O–Fe–O [34]. The aforementioned results in relation to the EDS analysis (Figure 3) clearly denote the presence of iron oxide on the surface of the silica microspheres.

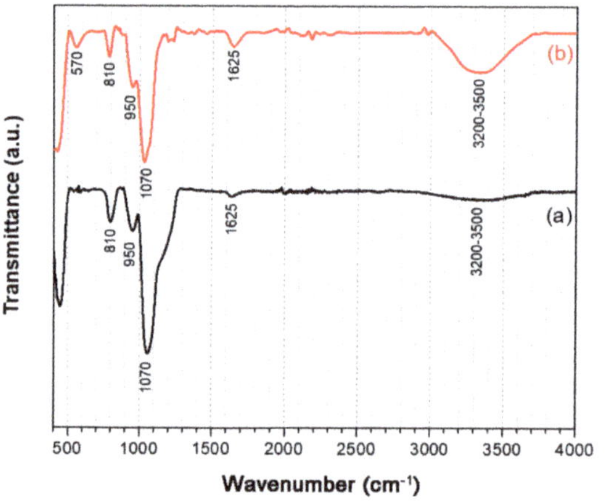

Figure 7. FT-IR spectra of (**a**) pristine silica microspheres and (**b**) Fe^+—functionalized silica microspheres.

3.3. *Microwave Absorption Evaluation*

The triggerable properties of the produced microparticles were estimated via microwave absorption evaluation. Figure 8 illustrates the samples tested (top) and the results of the microwave absorption and the subsequent heat production of each measured sample, as identified by the IR thermo-camera (bottom). Four measurements took place: (a) an empty vial as a control, (b) silica microspheres, (c) pristine commercial MWCNTs and (d) SiO_2@CNTs. Each sample was microwave irradiated for 2 min, and the temperature was measured inside the oven directly after microwave irradiation came to a halt; subsequently, the oven was left to cool for 5 min before starting the next measurement. On the top left of each image, the measured temperature is highlighted, whereas on the right side of each image, the temperature scale bar is presented. The bright yellow-red area observed in the background, which is present in all pictures, is the area directly below the microwave source.

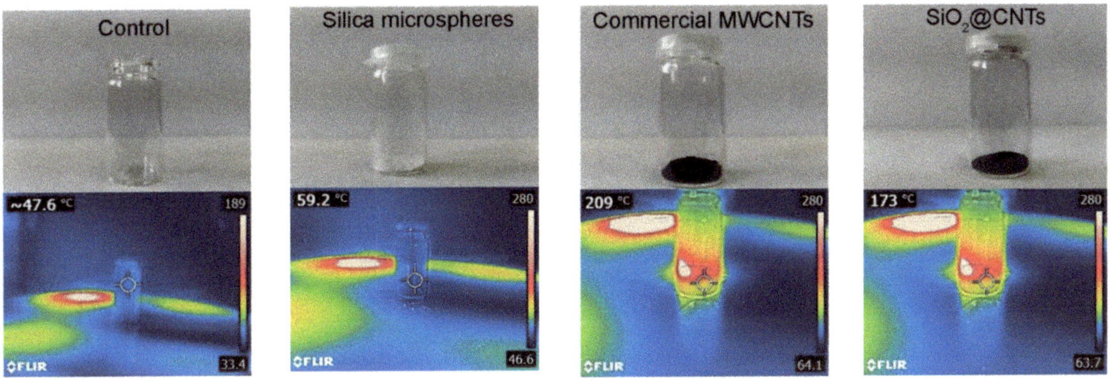

(**a**) (**b**) (**c**) (**d**)

Figure 8. Photographs of the samples prior to (**top**) and after their microwave irradiation (**bottom**): (**a**) an empty vial as a control; (**b**) silica microspheres; (**c**) pristine commercial MWCNTs; (**d**) SiO_2@CNTs.

Taking into account the obtained results, it can be observed that the silica spheres exhibit a slightly increased temperature (59.2 °C) compared to the control sample, whereas the pristine commercial MWCNTs and the SiO_2@CNTs exhibit elevated temperatures, 209 °C and 173 °C, respectively. Furthermore, it can be seen that there is an increase in the temperature scale bar limits, from the control to the rest of the samples, from 33.4–189 °C up to 64–280 °C, indicating that the silica microspheres, pristine commercial MWCNTs and the hybrid SiO_2@CNTs contribute to heat production. On the other hand, it may be remarked that despite the fact that the SiO_2@CNTs contain the same amount of MWCNTs as the pristine commercial MWCNTs, there is a difference in the observed temperature (~30 °C). This variance in temperature can be attributed to the difference between the I_d/I_g ratios of the MWCNTs, as observed in the Raman analysis (Figure 5). It has been previously reported that morphological factors such as structural defects, aspect ratio, specific surface area and purity are prevailing factors that dictate the microwave absorption of carbon nanotubes [35]. In addition, considering that defect-free SWCNTs have been previously microwave irradiated in a similar manner and their heat production was significantly higher, the above hypothesis can be further supported [10].

3.4. The Outcome Analysis

It can be seen that three different materials were used. Taking the literature into account, in a previous study, Argawal et al. utilized the floating catalyst approach (xylene/ferrocene mixture) to grow MWCNTs on silica microspheres of various sizes [36]. Their results demonstrated that the size of the silica microspheres has an impact on MWCNTs' growth, whereas smaller values of the I_d/I_d ratio were obtained for microspheres with sizes larger than 490 nm. In addition, Zhou et al. functionalized silica nanoparticles with iron chloride, using methane as a carbon source, and produced SWCNTs; for silica spheres larger than 500 nm, the SWCNTs spanned across different silica spheres, as also described in our work (Figure 1), whereas for larger silica spheres, SWCNTs engulfed the surface of their individual silica microspheres [37]. Nevertheless, Wei et al. modified silicon wafers with planar silica micro-discs and produced MWCNTs with ordered, preferential growth on the surface of the silica discs, utilizing the floating catalyst approach (xylene/ferrocene mixture) [38].

Evidently, the obtained results in the presented work agree with the above-mentioned research outcomes, further supporting the hypothesis that the growth of CNTs on silica microparticle substrates is affected by the size and the available surface area of the substrate. Silica microparticle substrates larger than 500 nm produce MWCNTs with fewer structural defects (judging by the I_d/I_g ratio), whereas the appropriate selection of the carbon source and catalyst, with either a floating catalyst (xylene/ferrocene mixture) or a supported catalyst (Fe^+—functionalized silica microspheres with methane or acetylene as a carbon source), yields CNTs with ordered and random growth, respectively. In our case, the functionalization conditions, such as the weight percentage of the silica microspheres and the iron chloride concentration in the functionalization medium, were crucial for the functionalization of the entire population of silica microspheres and the homogenous growth of MWCNTs on each microparticle. Table 1 tabulates the physical and chemical properties of the three different materials that were used in our work.

Table 1. Tabulated values of physical and chemical properties of the used materials.

	Commercial MWCNTs	SiO_2	SiO_2@CNTs
Physical Properties			
Size (nm)	25–35 (tube diameter)	350 (sphere)	-
Zeta potential (mV)	-	−40	-
Thermal degradation (°C)	450–600 [39]	stable	463–670
I_d/I_g	0.6	-	2.0
Heat production (°C)	209.0	59.2	173.0
Chemical Properties			
Surface functionalization	inert	Fe; –OH	–OH
Composition (wt.%)	94.1–C; 5.9–O	49.2–O; 48.0–Si; 2.8–Fe	37.0–C; 31.5–O; 30.3–Si; 1.2–Fe

Regarding the microwave absorption of similar materials, according to the literature, the study of Hekmatara et al. investigated the microwave absorption properties of Fe_3O_4/MWCNT materials coated with silica. The results revealed increased microwave absorption compared to the uncoated sample (Fe_3O_4/MWCNT) [40]. In the work of Xiang et al., MWCNT-fused silica composites were fabricated, demonstrating enhanced microwave attenuation properties [41]. Barron and his team performed experiments on the modification of ceramic particle surfaces with carbon nanotubes and silica. The exposure of the produced hybrid materials to a microwave source resulted in heating the carbon nanotubes [42]. In another study, Yuan et al. evaluated the microwave attenuation performances of SiO_2-MWCNTs matrix composites. It was proved that the obtained materials had increased microwave absorption [43].

Taking into account the aforementioned results regarding microwave absorption, it may be remarked that the comparison of our results with the literature could possibly lead to misinterpretation. The reason is that, in our study, the applied method for microwave absorption was used to compare the hybrid material (SiO_2@CNTs) with the reference (MWCNTs) in order to approximately estimate the application performance.

4. Conclusions

Silica microspheres functionalized with iron chloride were fabricated and used for the growth of MWCNTs on their surface. The obtained hybrid structures consisted of 80 wt.% silica and 20 wt.% MWCNTs, with an average diameter of 50 nm and an average degree of structural defects. The SiO_2@CNTs structures after 2 min microwave irradiation (magnetron 700 W operating at 2.45 GHz) produced enough heat to increase the material's temperature from ambient temperature up to 173 °C, compared to the pristine commercial MWCNTs, which exhibited a measured temperature of 209 °C. By correlating these results with similar research in the literature related to the growth of CNTs on silica microspheres, it can be assumed that the size of the silica microspheres, the functionalization process with iron chloride and the selected carbon source are key variables for the physical characteristics of the resulting MWCNTs in terms of morphology and structural defects. Evidently, these hybrid structures can be tailored by modulating the aforementioned variables in order to produce structures with specific microwave absorption potential, whereas the silica part has the potential to be further functionalized in order to provide chemical affinity with polymer resins, contrary to the chemical inertness of pristine MWCNTs. The results from this study could greatly benefit polymer technology and the coating industry.

Author Contributions: Conceptualization, P.K. and I.A.K.; methodology, P.K. and I.A.K.; software, P.K. and I.A.K.; validation, P.K., I.A.K. and C.A.C.; formal analysis, P.K. and I.A.K.; investigation, P.K. and I.A.K.; resources, P.K. and I.A.K.; data curation, P.K.; writing—original draft preparation, P.K. and I.A.K.; writing—review and editing, I.A.K. and C.A.C.; visualization, P.K. and I.A.K.; supervision, C.A.C.; project administration, C.A.C.; funding acquisition, C.A.C. All authors have read and agreed to the published version of the manuscript.

Funding: This research was funded by the HORIZON 2020 Collaborative project "DECOAT" (Recycling of coated and painted textile and plastic materials, Grant agreement no.: 814505).

Institutional Review Board Statement: Not applicable.

Informed Consent Statement: Not applicable.

Data Availability Statement: Data sharing not applicable.

Conflicts of Interest: The authors declare no conflict of interest.

References

1. Rhein, S.; Sträter, K.F. Corporate self-commitments to mitigate the global plastic crisis: Recycling rather than reduction and reuse. *J. Clean. Prod.* **2021**, *296*, 126571. [CrossRef]
2. Wypych, G. 3-mechanisms of adhesion loss. In *Handbook of Adhesion Promoters*; Wypych, G., Ed.; ChemTec Publishing: Scarborough, ON, Canada, 2018; pp. 45–53. [CrossRef]
3. Bandl, C.; Kern, W.; Schlögl, S. Adhesives for "debonding-on-demand": Triggered release mechanisms and typical applications. *Int. J. Adhes. Adhes.* **2020**, *99*, 102585. [CrossRef]
4. Banea, M. Debonding on Demand of Adhesively Bonded Joints: A Critical Review. *Rev. Adhes. Adhes.* **2019**, *7*, 33–50. [CrossRef]
5. Ferahian, A.-C.; Hohl, D.K.; Weder, C.; Montero de Espinosa, L. Bonding and Debonding on Demand with Temperature and Light Responsive Supramolecular Polymers. *Macromol. Mater. Eng.* **2019**, *304*, 1900161. [CrossRef]
6. Guo, D.; Xie, G.; Luo, J. Mechanical properties of nanoparticles: Basics and applications. *J. Phys. D Appl. Phys.* **2014**, *47*, 013001. [CrossRef]
7. Kartsonakis, I.A.; Goulis, P.; Charitidis, C.A. Triggerable Super Absorbent Polymers for Coating Debonding Applications. *Polymers* **2021**, *13*, 1432. [CrossRef] [PubMed]
8. Goulis, P.; Kartsonakis, I.A.; Charitidis, C.A. Synthesis and Characterization of a Core-Shell Copolymer with Different Glass Transition Temperatures. *Fibers* **2020**, *8*, 71. [CrossRef]
9. Gupta, N.; Gupta, S.M.; Sharma, S.K. Carbon nanotubes: Synthesis, properties and engineering applications. *Carbon Lett.* **2019**, *29*, 419–447. [CrossRef]
10. Imholt, T.J.; Dyke, C.A.; Hasslacher, B.; Perez, J.M.; Price, D.W.; Roberts, J.A.; Scott, J.B.; Wadhawan, A.; Ye, Z.; Tour, J.M. Nanotubes in Microwave Fields: Light Emission, Intense Heat, Outgassing, and Reconstruction. *Chem. Mater.* **2003**, *15*, 3969–3970. [CrossRef]
11. Strozzi, M.; Pellicano, F. Linear vibrations of triple-walled carbon nanotubes. *Math. Mech. Solids* **2017**, *23*, 1456–1481. [CrossRef]
12. He, X.Q.; Eisenberger, M.; Liew, K.M. The effect of van der Waals interaction modeling on the vibration characteristics of multiwalled carbon nanotubes. *J. Appl. Phys.* **2006**, *100*, 124317. [CrossRef]
13. Pantano, A.; Parks, D.M.; Boyce, M.C. Mechanics of deformation of single- and multi-wall carbon nanotubes. *J. Mech. Phys. Solids* **2004**, *52*, 789–821. [CrossRef]
14. Vázquez, E.; Prato, M. Carbon Nanotubes and Microwaves: Interactions, Responses, and Applications. *ACS Nano* **2009**, *3*, 3819–3824. [CrossRef]
15. Wadhawan, A.; Garrett, D.; Perez, J.M. Nanoparticle-assisted microwave absorption by single-wall carbon nanotubes. *Appl. Phys. Lett.* **2003**, *83*, 2683–2685. [CrossRef]
16. Tobias, G.; Mendoza, E.; Ballesteros, B. Functionalization of Carbon Nanotubes. In *Encyclopedia of Nanotechnology*; Bhushan, B., Ed.; Springer: Dordrecht, The Netherlands, 2012; pp. 911–919. [CrossRef]
17. Stöber, W.; Fink, A.; Bohn, E. Controlled growth of monodisperse silica spheres in the micron size range. *J. Colloid Interface Sci.* **1968**, *26*, 62–69. [CrossRef]
18. Wang, X.; Huang, J.; Feng, H.; Li, J.; Pu, Z.; Yin, X. Facile preparation of the dendritic Fe_3O_4 with a core-shell microstructure in SiO_2-B_2O_3-Al_2O_3-CaO-Fe_2O_3 glass-ceramic system for enhanced microwave absorbing performance. *J. Alloys Compd.* **2021**, *877*, 160147. [CrossRef]
19. Green, M.; Liu, Z.; Xiang, P.; Liu, Y.; Zhou, M.; Tan, X.; Huang, F.; Liu, L.; Chen, X. Doped, conductive SiO_2 nanoparticles for large microwave absorption. *Light Sci. Appl.* **2018**, *7*, 87. [CrossRef]

20. Kainourgios, P.; Tziveleka, L.-A.; Kartsonakis, I.A.; Ioannou, E.; Roussis, V.; Charitidis, C.A. Silver Nanoparticles Grown on Cross-Linked Poly (Methacrylic Acid) Microspheres: Synthesis, Characterization, and Antifungal Activity Evaluation. *Chemosensors* **2021**, *9*, 152. [CrossRef]
21. Yu, Z.; Chen, D.; Tøtdal, B.; Holmen, A. Effect of catalyst preparation on the carbon nanotube growth rate. *Catal. Today* **2005**, *100*, 261–267. [CrossRef]
22. Yue, R.; Chen, Q.; Li, S.; Zhang, X.; Huang, Y.; Feng, P. One-step synthesis of 1,6-hexanediamine modified magnetic chitosan microspheres for fast and efficient removal of toxic hexavalent chromium. *Sci. Rep.* **2018**, *8*, 11024. [CrossRef]
23. Chiani, E.; Azizi, S.N.; Ghasemi, S. PdCu bimetallic nanoparticles decorated on ordered mesoporous silica (SBA-15)/MWCNTs as superior electrocatalyst for hydrogen evolution reaction. *Int. J. Hydrog. Energy* **2021**, *46*, 25468–25485. [CrossRef]
24. Mansfield, E.; Kar, A.; Hooker, S.A. Applications of TGA in quality control of SWCNTs. *Anal. Bioanal. Chem.* **2010**, *396*, 1071–1077. [CrossRef] [PubMed]
25. Javed, M.; Abbas, S.M.; Hussain, S.; Siddiq, M.; Han, D.; Niu, L. Amino-functionalized silica anchored to multiwall carbon nanotubes as hybrid electrode material for supercapacitors. *Mater. Sci. Energy Technol.* **2018**, *1*, 70–76. [CrossRef]
26. Yudianti, R.; Indrarti, L.; Onggo, H. Thermal Behavior of Purified Multi Walled Carbon Nanotube. *J. Appl. Sci.* **2010**, *10*, 1978–1982. [CrossRef]
27. Silva, M.A.; Felgueiras, H.P.; de Amorim, M.T.P. Carbon based membranes with modified properties: Thermal, morphological, mechanical and antimicrobial. *Cellulose* **2019**, *27*, 1497–1516. [CrossRef]
28. Maddalena, R.; Hall, C.; Hamilton, A. Effect of silica particle size on the formation of calcium silicate hydrate [C-S-H] using thermal analysis. *Thermochim. Acta* **2019**, *672*, 142–149. [CrossRef]
29. Li, Z.Q.; Lu, C.J.; Xia, Z.P.; Zhou, Y.; Luo, Z. X-ray diffraction patterns of graphite and turbostratic carbon. *Carbon* **2007**, *45*, 1686–1695. [CrossRef]
30. Das, R.; Hamid, S.; Ali, M.; Ramakrishna, S.; Yongzhi, W. Carbon Nanotubes Characterization by X-ray Powder Diffraction—A Review. *Curr. Nanosci.* **2014**, *11*, 23–35. [CrossRef]
31. Iijima, S.; Ichihashi, T. Single-shell carbon nanotubes of 1-nm diameter. *Nature* **1993**, *363*, 603–605. [CrossRef]
32. Nordgreen, T.; Liliedahl, T.; Sjöström, K. Elemental Iron as a Tar Breakdown Catalyst in Conjunction with Atmospheric Fluidized Bed Gasification of Biomass: A Thermodynamic Study. *Energy Fuels* **2006**, *20*, 890–895. [CrossRef]
33. Azarshin, S.; Moghadasi, J.; Aboosadi, Z.A. Surface functionalization of silica nanoparticles to improve the performance of water flooding in oil wet reservoirs. *Energy Explor. Exploit.* **2017**, *35*, 685–697. [CrossRef]
34. Ahangaran, F.; Hassanzadeh, A.; Nouri, S. Surface modification of Fe_3O_4@SiO_2 microsphere by silane coupling agent. *Int. Nano Lett.* **2013**, *3*, 1–5. [CrossRef]
35. Savi, P.; Giorcelli, M.; Quaranta, S. Multi-Walled Carbon Nanotubes Composites for Microwave Absorbing Applications. *Appl. Sci.* **2019**, *9*, 851. [CrossRef]
36. Agrawal, S.; Kumar, A.; Frederick, M.J.; Ramanath, G. Hybrid Microstructures from Aligned Carbon Nanotubes and Silica Particles. *Small* **2005**, *1*, 823–826. [CrossRef]
37. Zhou, W.; Zhang, Y.; Li, X.; Yuan, S.; Jin, Z.; Xu, J.; Li, Y. Preferential Growth of Single-Walled Carbon Nanotubes on Silica Spheres by Chemical Vapor Deposition. *J. Phys. Chem. B* **2005**, *109*, 6963–6967. [CrossRef]
38. Wei, B.Q.; Vajtai, R.; Jung, Y.; Ward, J.; Zhang, R.; Ramanath, G.; Ajayan, P.M. Organized assembly of carbon nanotubes. *Nature* **2002**, *416*, 495–496. [CrossRef]
39. Mahajan, A.; Kingon, A.; Kukovecz, Á.; Konya, Z.; Vilarinho, P.M. Studies on the thermal decomposition of multiwall carbon nanotubes under different atmospheres. *Mater. Lett.* **2013**, *90*, 165–168. [CrossRef]
40. Hekmatara, H.; Seifi, M.; Forooraghi, K.; Mirzaee, S. Synthesis and microwave absorption characterization of SiO_2 coated Fe_3O_4-MWCNT composites. *Phys. Chem. Chem. Phys. PCCP* **2014**, *16*, 24069–24075. [CrossRef]
41. Xiang, C.; Pan, Y.; Liu, X.; Sun, X.; Shi, X.; Guo, J. Microwave attenuation of multiwalled carbon nanotube-fused silica composites. *Appl. Phys. Lett.* **2005**, *87*, 123103. [CrossRef]
42. Gomez, V.; Dunnill, C.W.; Barron, A.R. A microwave cured flux for the adhesion of ceramic particles using silica coated carbon nanotubes. *Carbon* **2015**, *93*, 774–781. [CrossRef]
43. Wen, B.; Cao, M.-S.; Hou, Z.-L.; Song, W.-L.; Zhang, L.; Lu, M.-M.; Jin, H.-B.; Fang, X.-Y.; Wang, W.-Z.; Yuan, J. Temperature dependent microwave attenuation behavior for carbon-nanotube/silica composites. *Carbon* **2013**, *65*, 124–139. [CrossRef]

fibers

MDPI

Article

Synthesis and Characterization of a Core-Shell Copolymer with Different Glass Transition Temperatures

Panagiotis Goulis, Ioannis A. Kartsonakis * and Costas A. Charitidis *

Research Unit of Advanced, Composite, Nano-Materials and Nanotechnology, School of Chemical Engineering, National Technical University of Athens, 9 Heroon Polytechniou St., Zographos, GR-15773 Athens, Greece; pgoulis@chemeng.ntua.gr

* Correspondence: ikartso@chemeng.ntua.gr (I.A.K.); charitidis@chemeng.ntua.gr (C.A.C.); Tel.: +30-21-0772-4046 (C.A.C.)

Received: 4 November 2020; Accepted: 17 November 2020; Published: 23 November 2020

check for updates

Abstract: The aim of this study is to synthesize an organic core-shell co-polymer with a different glass transition temperature (T_g) between the core and the shell that can be used for several applications such as the selective debonding of coatings or the release of encapsulated materials. The co-polymer was synthesized using free radical polymerization and was characterized with respect to its morphology, composition and thermal behavior. The obtained results confirmed the successful synthesis of the co-polymer copolymer poly(methyl methacrylate)@poly(methacrylic acid-co-ethylene glycol dimethacrylate), *PMMA@P(MAA-co-EGDMA)*, which can be used along with water-based solvents. Furthermore, the T_g of the polymer's core *PMMA* was 104 °C, while the T_g of the shell *P(MAA-co-EGDMA)* was 228 °C, making it appropriate for a wide variety of applications. It is worth mentioning that by following this specific experimental procedure, methacrylic acid was copolymerized in water, as the shell of the copolymer, without forming a gel-like structure (hydrogel), as happens when a monomer is polymerized in aqueous media, such as in the case of super-absorbent polymers. Moreover, the addition and subsequent polymerization of the monomer methyl methacrylate (*MAA*) into the mixture of the already polymerized *PMMA* resulted in a material that was uniform in size, without any agglomerations or sediments.

Keywords: core-shell; polymers; glass temperature; SEM; TEM

1. Introduction

In recent years, the need to synthesize new core-shell polymers has increased as chemical procedures have required accuracy and effectiveness more than ever. Micro- and nanostructured core-shell copolymers solve many problems of materials industry, medicine and infrastructure, while remaining practically easy to synthesize and apply [1].

Core-shell polymers can be manufactured through a vast number of experimental means, including various methods such as electrospinning [2] and solvent-free and large scale processes [3], as well as ultrasound-assisted methods [4]. Furthermore, material properties have been studied under several circumstances and environmental conditions, such as laser fields, mini-emulsions, pre-emulsions and acid media [5,6].

Many attempts have been made to synthesize a novel core-shell in the last five years. In order to cover the continuously increasing number of modern needs for core-shells, scientists have tried many new chemistry-based approaches, while attempting to keep the total cost relatively low. A few novel approaches from the literature include Hu et al., who wrote a review about a single-component core-shell [7]; Han et al., who developed a core-shell magnetic material with applications in food chemistry [8]; Gul et al., who managed to imprint a polymer onto a core-shell material with magnetic properties [9]; and Farboudi et al., who synthesized core-shell nanofibers grafted with chitosan for the controlled release of drugs [10]. Furthermore, several studies were conducted by our group based on the production of core-shell particles and microcapsules that can be used either for the removal of chlorides from water or as additives into coatings that can protect metal alloys from corrosion [11].

In recent years, research focusing on this type of polymer has increased. Core-shell polymers have been used for numerous biological and medical applications, such as controlled calcium and drug delivery, antibacterial applications, stem cells, wound healing, virus chromatography, cancer and hyperthermia therapy and glaucoma treatment [12–14]. They have also been used for electronics and energy applications such as solar cells, energy storage, batteries, photovoltaics, fuel cells and sensors, as they enhance electrical conductivity and electrical properties in general [15–20]. However, the majority of core-shell materials concern engineering applications such as mechanical reinforcements, where they are used as fillers for their high strength and tensile performance [21] as well as for their high interfacial strength between the cell and the core [22]. They are also used for toughening epoxy resins [23].

Additionally, lots of synthetic processes and characterization techniques have been deployed in order to investigate core-shell materials more accurately and fully understand their effects on other compounds. Various methods are already known, with in situ synthesis, emulsion polymerization, catalysis, use of particles as sacrificial templates, and use of reactive surfactants being some of the most popular among them research in this field remains ongoing [24–27]. Moreover, acidic media, transmission electron microscopy (TEM) and other numerous techniques have shed light on the structure, morphology and characteristics of these materials [28–31].

In this study, the concept was to synthesize core-shell polymers that could be triggered by heat. Therefore, the shell can be selectively weakened by an external trigger (heat), while leaving the core exposed, to release its encapsulations at a specific moment in a given targeted area. For this to happen, it is necessary that the polymers that form the core and the cell have different glass transition temperatures (T_g), meaning that the T_g of the shell should be higher than the T_g of the core [32]. According to this idea, a core-shell polymer (core@shell) was produced—poly(methyl methacrylate)@poly(methacrylic acid), *PMMA@P(MAA-co-EGDMA)*—in which the T_g of the core and the shell were 104 and 228 °C, respectively. This organic polymer was characterized via scanning electron microscopy (SEM), Fourier-transform infrared spectroscopy (FT-IR), TEM, thermogravimetric analysis (TGA) and differential scanning calorimetry (DSC). It should be mentioned, that, following this experimental path, *MAA* was polymerized via classic radical polymerization using water as solvent, without forming a hydrogel. The produced material was uniform in size, without any agglomerations or sediments.

Finally, it may be remarked that the added value of our work resides in the fact that this copolymer combined hydrophilic and lipophilic parts and different glass transition temperatures, while it was synthesized in a one-pot synthetic process.

2. Materials and Methods

2.1. Materials

Analytical reagent grade chemicals were used. Methacrylic acid (MAA, Fisher Chemicals, Waltham, MA, USA), methyl methacrylate (MMA, Sigma Aldrich, St. Louis, Missouri, USA), ethylene glycol dimethacrylate (EGDMA, Sigma Aldrich, St. Louis, MO, USA) and potassium peroxodisulphate (KPS, Acros Organics, Geel, Belgium) were used as received without further purification. Also, distilled water was used. Methacrylic acid and methyl methacrylate were distilled under vacuum prior to use.

2.2. Synthesis of Core-Shell Copolymer

Radical polymerization was performed for the synthesis of the thermo-responsive core-shell co-polymer *PMMA@(PMAA-co-EGDMA)*. First, 20 mL of methyl methacrylate was dissolved in 500 mL of water and nitrogen gas was flushed into a round, three-neck, 1 L glass reaction flask. Then, 0.2 g potassium persulfate (KPS-initiator) was added and the temperature was adjusted to 90 °C for the initiation of the polymerization. A glass reflux condenser was used to recycle the solvent during the polymerization reaction. After 4 h, when the polymerization of methyl methacrylate was completed, 20 mL of distilled methacrylic acid was added to the flask together with 2 mL of crosslinker (*EGDMA*) and 0.2 g of initiator (KPS). The reaction was left to react for 4 h under vigorous stirring at 90 °C. After this, the contents of the flask were centrifuged and gently dried in an oven at 40 °C for 24 h. Figure 1 illustrates the schema of the produced core-shell copolymer *PMMA@P(MAA-co-EGDMA)*. The produced thermo-responsive core-shell nanostructured co-polymer was characterized in respect to its morphology, composition and structure.

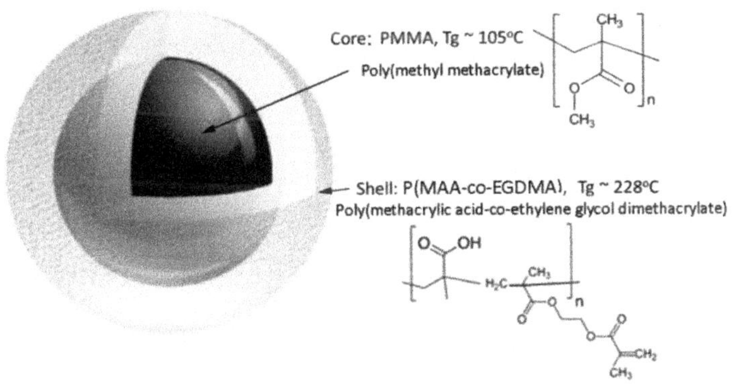

Figure 1. Schematic representation of the produced core-shell copolymer *PMMA@P(MAA-co-EGDMA)*.

2.3. Characterization

The morphology and composition of the produced materials were estimated with an ultra-high resolution scanning electron microscope (UHR-SEM) using NOVA NANOSEM 230 (FEI Company) coupled with a Hitachi electron microscope TM3030 and an energy dispersive X-ray spectrophotometer (EDS) (QUANTAX 70) [33]. In addition, the morphology of the core-shell copolymers was also characterized by TEM (model FEI CM20), operated at an accelerating voltage of 200 kV [34–36].

The FT-IR spectra of the produced polymers was analyzed via an attenuated total reflectance FT-IR instrument, the Cary 630 spectrometer (Agilent), with a resolution of 4 cm^{-1} and an operating wavelength range of 4000–400 cm^{-1} [34]. Thermal analysis of the produced polymers was conducted via a TGA/DSC instrument, the STA 449 F5 Jupiter. The system consists of an SiC furnace with an operation temperature varying between 25 and 1550 °C. The gas used was nitrogen at a volumetric rate of 50 mL/min and at a heating rate of 20 °C/min [34]. The size distribution of the produced polymers was estimated via dynamic light scattering (DLS) measurements using a Zetasizer Nano ZS.

Gel permeated chromatography (GPC) analysis was performed using a modular instrument consisting of a waters model 510 pump, a U6K sample injector, a 401 refractometer and a set of four 1 μm-Styragel columns with a continuous porosity range of 10^6–10^3 Å. The carrier solvent was $CHCl_3$ at a flow rate of 1 mL/min. The system was calibrated with seven polystyrene (PS) standards having molecular weights (MWs) between 1000 and 900,000 g/mol. The system was operated at 25 °C.

3. Results and Discussion

3.1. Morphological Characterization (SEM and TEM)

Figure 2a illustrates the SEM image of the produced core-shell copolymer *PMMA@P(MAA-co-EGDMA)*. It can be seen that the synthetic process of producing the material resulted in a copolymer with a spherical morphology. Taking into account Figure 2b, which depicts a corresponding particle size distribution histogram with values from 63 core-shell copolymer particles, it can be seen that the diameter of these spheres was in the nanoscale, ranging from approximately 160 to 210 nm [37]. These properties indicate that this material's volume was appropriate for its use in several applications that would require the capsules to take up as little space as possible (e.g., incorporation into coatings or layers). It can be seen from the transmission electron microscopy images (Figure 3) that there was a thin shell on the outside of the polymeric core [38,39].

Figure 2. SEM image (**a**) and particle size distribution histogram (**b**) of the *PMMA@P(MAA-co-EGDMA)* core-shell copolymer.

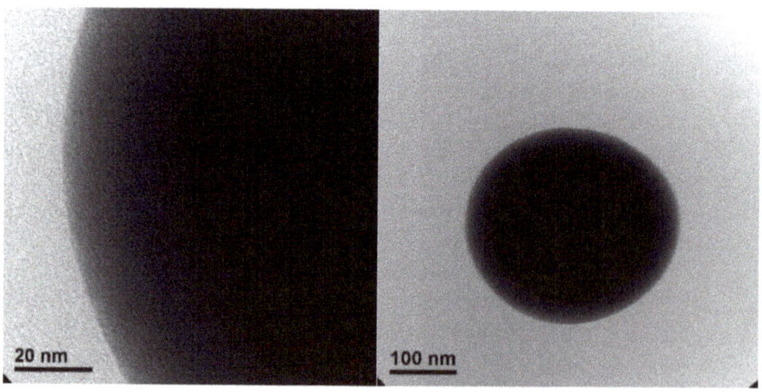

Figure 3. TEM images of the core-shell copolymer *PMMA@P(MAA-co-EGDMA)*.

The thin shell of *P(MAA-co-EGDMA)* around the core of *PMMA* protects and surrounds the core, while on the other hand it can easily be weakened and removed without the use of a great amount of energy, if an application requires the core to be exposed relatively easy for the selective release of ingredients encapsulated in the core (e.g., targeted delivery or coatings).

3.2. Thermal Analysis (TGA and DSC)

Thermogravimetric Analysis was performed in inert atmosphere with a gas flow of 50 mL/min and a heating rate of 5 °C/min. Figure 4 illustrates the TGA/DSC graphs of the produced (a) *PMMA* core and (b) *PMMA@P(MAA-co-EGDMA)* core-shell copolymer. From the TGA curve of Figure 4a, it can be seen that the weight loss of the *PMMA* (core) was much more abrupt than in the case of the core-shell copolymer. This was expected, because the presence of the shell in the second case allowed a slower degradation, which was also smoother. The slight increase in mass at the start of the measurement is attributable to the scale drift of the instrument. This effect occurs when the measurement starts after the temperature has been increased from room temperature to 100 °C at an accelerating rate.

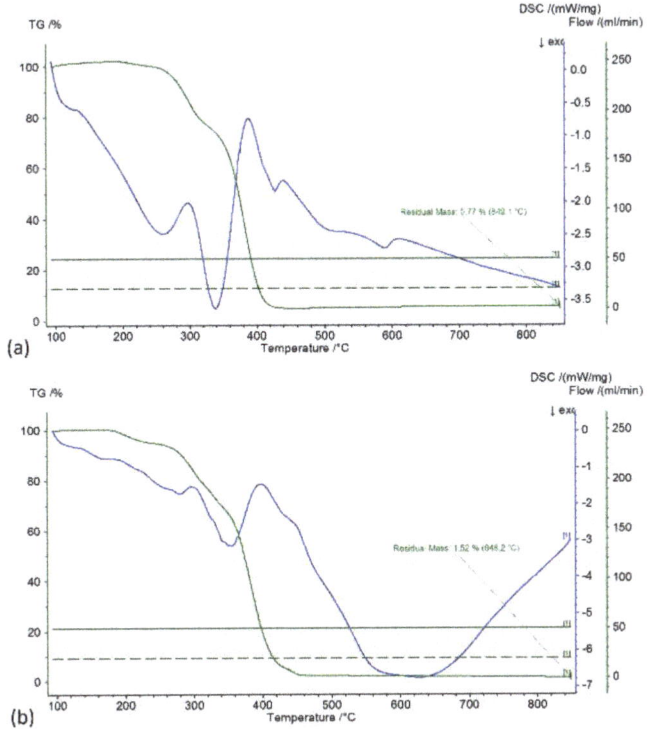

Figure 4. TGA/DSC graphs of the produced *PMMA* core (**a**) and *PMMA@P(MAA-co-EGDMA)* core-shell copolymer (**b**).

The DSC endothermic curve at 400 °C on the same graph corresponds to the polymer's degradation. The exothermic curve at 340 °C can be attributed to the C-OCH_3 bond rupture, which enriched the nitrogen atmosphere with oxygen and led to partial oxidation of the polymer [40]. From the TGA curve of the Figure 4b, it can be readily observed that there was a first weight loss between 200 and 290 °C that can be attributed to the degradation of *PMMA*, and the second sharp weight loss between 290 and 440 °C can be attributed to the degradation of *P(MAA-co-EGDMA)*. Furthermore, the DSC curve (blue) demonstrates two endothermic curves, the first small one at 280 °C and the second big one at 360 °C, which correspond to the two thermal decompositions of *PMMA* and *P(MAA-co-EGDMA)*, respectively. The residual mass at 850 °C is the remaining carbon, approximately 1.5% of the initial mass [41].

Finally, it should be mentioned that an exothermic DSC curve can be observed in the TGA/DSC graphs of Figure 4a at about 105 °C, where the T_g of the *PMMA* is 104 °C. Additionally, in the TGA/DSC graphs of Figure 4b, there is an exothermic DSC curve at about 230 °C where the T_g of the *PMAA* is 228 °C. These two curves correspond to the respective T_g of the two polymers, although their intensity was low, due to the slight decrease in bond strength after the T_g point. On the other hand, the DSC exothermic curves corresponding to the decomposition had high intensity.

3.3. *Attenuated Total Reflectance Fourier-Transform Infrared Spectroscopy (FT-IR)*

The FT-IR graphs concerning the *PMMA* core as well as the *PMMA@P(MAA-co-EGDMA)* core-shell copolymer are illustrated in Figure 5 and the corresponding peaks are tabulated in Table 1. The peak at 1156 cm^{-1} corresponds to the -CH_2- bond wagging of the aliphatic chains of both *PMMA* and *PMAA*. Moreover, the peak at 475 cm^{-1} corresponds to the C-C=O in the plane deformation vibration of the carboxyls of the two polymers, while the peaks at 747 and 840 cm^{-1} can be ascribed to the C-H deformation vibration and the peak at 963 cm^{-1} is related to the C-O stretching vibration. The peak at 1187 cm^{-1} can be attributed to the -OCH_3 vibration (methyl methacrylate) [42]. Furthermore, the peaks at 1237, 1384 and 1435 cm^{-1} can be ascribed to the -CH- bending of the methyl groups, which were present in both compounds, as well as to the -CH_3 deformation of the branches of the main polymer chain [43]. A C-H scissor vibration can be seen at 1474 cm^{-1}, while the C=C stretching vibration can be observed at 1636 cm^{-1}. Additionally, the -C=O stretch can be noticed at 1716 cm^{-1}. The stretching of -CH_3 is visible at 2925 cm^{-1}, while the stretching of -CH_2 can be seen at 2996 cm^{-1} [44]. Finally, the -O-H stretches of methacrylic acid can be seen in the region of 3400 cm^{-1} [45].

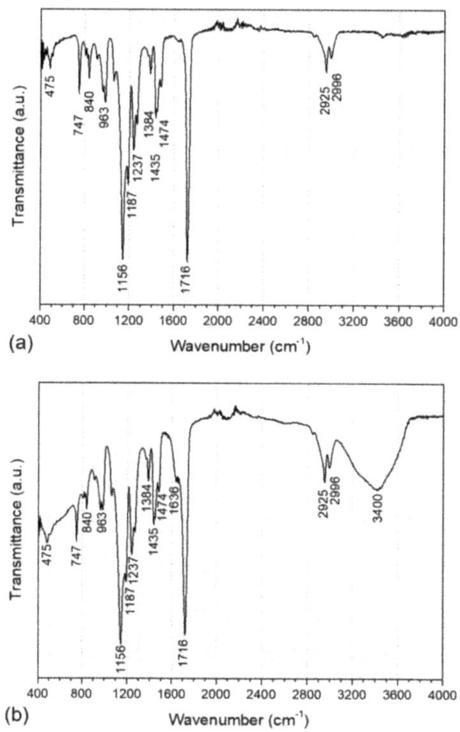

Figure 5. FT-IR spectra of the *PMMA* core polymer (**a**) and *PMMA@P(MAA-co-EGDMA)* core-shell copolymer (**b**).

Table 1. FT-IR peak analysis.

Wavenumber (cm^{-1})	Corresponding Bond
475	C-C=O in plane deformation vibration
747, 840	C-H deformation vibration
963	C-O stretching vibration
1156	-CH_2- wagging and C-O phenol stretch
1187	-OCH_3 vibration (methyl methacrylate)
1237, 1384, 1435	C-H and O-H deformation vibration
1474	C-H scissor vibration
1636	C=C stretching vibration (EGDMA)
1716	-C=O stretch
2925	-CH_3 stretching
2996	-CH_2 stretching
3400	-O-H stretch (methacrylic acid)

3.4. Dynamic Light Scattering (DLS)

Dynamic light scattering (DLS) analysis is very useful for polymer characterization, because it can offer important information about organic compounds. DLS can be used to monitor polymer formation processes, in diffusion analysis and to characterize polymer solutions and surfactants, as well as to investigate natural polymers [46]. Moreover, a lot of properties can be determined via DLS analysis such as viscosity, polymerization mechanisms, molecular weight and molecular size, even particle size and particle size distribution. The study of reaction kinetics during polymerization and the detection of polymeric structures in emulsions are also feasible using DLS [47]. In our case, about 10 mg of the solid polymer was dispersed in 5 mL of water in a small vial. The vial was left under supersonic vibrations for 10 min so that the polymer would be sufficiently dispersed for DLS characterization. The Z-Average diameter was 446 nm, while the polydispersity index (PDI) was 0.077 (Figure 6). These facts denote that the sample was uniform in size distribution, without any agglomerations or sediments [48]. It should be mentioned that the average diameter of the produced *PMMA@P(MAA-co-EGDMA)* core-shell copolymer determined by DLS was higher than the values illustrated in the particle size distribution histogram of Figure 2b. The reason for this is that the hydrodynamic volume of the materials was measured as part of the DLS method. In our case, the produced *PMMA@P(MAA-co-EGDMA)* core-shell copolymer interacted with the solvent that was used for the measurement (water), resulting in an expansion of its hydrodynamic volume.

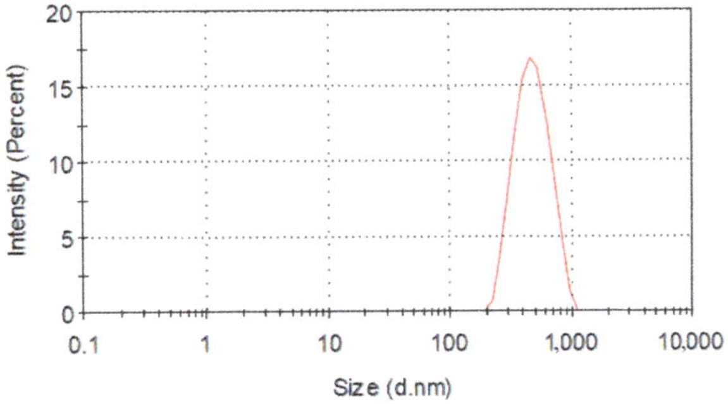

Figure 6. Particle size distribution by intensity of *PMMA@P(MAA-co-EGDMA)* core-shell copolymer.

3.5. Gel Permeated Chromatography

GPC was used to estimate the molecular weights of the produced materials. Unfortunately, it was not feasible to determine the molecular weight of the *PMMA@P(MAA-co-EGDMA)* core-shell copolymer due to solubility difficulties. In detail, GPC measurements were conducted using four different chromatography solvents including water with acetonitrile, chloroform, tetrahydrofuran, and dimethylformamide (DMF). The *PMMA* (core) was fully soluble in DMF, allowing a GPC graph to be obtained. On the other hand, the *PMMA@P(MAA-co-EGDMA)* core-shell copolymer was not soluble in any of these four solvents because it formed a gel-like two-phase structure that looked like an emulsion or suspension in the solvents. It should be mentioned that this final remark underlines the successful synthesis of the core-shell copolymer between the lipophilic *PMMA* and the hydrophilic *PMAA*, although it was not possible to obtain a molecular weight measurement. The distinctive ability of the instrument was up to a million Daltons. Judging from the GPC graph (Figure 7), it can be observed that the polydispersity index of the *PMMA* was relatively high, as there was a very wide distribution of molecular weights after the elution of the polymer (Table 2). The molecular weight ranged from one million to a few thousand Daltons. It should be noted that the first inverse peak (at an approximate elution time of 5 min) is attributable to the solvent and was not taken under consideration.

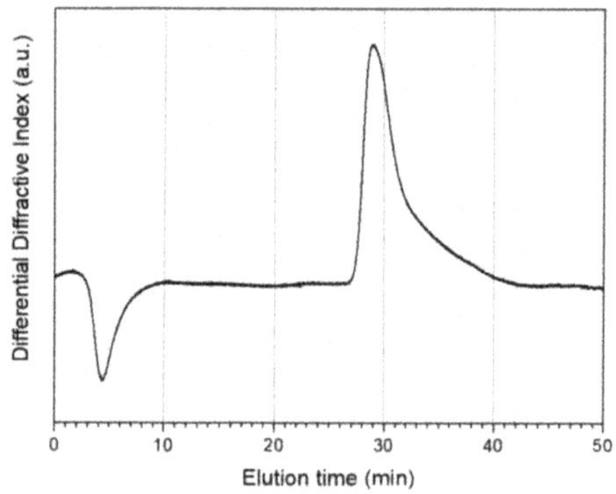

Figure 7. Gel permeated chromatography (GPC) traces of the *PMMA* (core) polymer.

Table 2. Tabulated values of the *PMMA* core polymer GPC measurement.

Retention Time (min)	Evolution Volume (mL)	Mol Wt (Daltons)	Calculated Weight (Daltons)
28.811	28.811	900,000	872,947
29.628	29.628	520,000	547,317
30.184	30.184	410,000	408,730
30.264	30.264	390,000	392,540
32.189	32.189	170,000	164,923
35.754	35.754	50,000	47,892
36.994	36.994	30,000	32,768
40.176	40.176	12,500	11,728
43.049	43.049	3350	3569
43.821	43.821	2500	2412

3.6. Outcome Analysis

The *Yield*(%) of the polymerization reaction for the synthesis of *PMMA@P(MAA-co-EGDMA)* core-shell copolymer can be calculated from the following Equations (1)–(3):

$$m_{MMA} = d_{MMA} \times V_{MMA} = 0.94\ {\scriptstyle g/mL} \times 20\ \text{mL} = 18.8\ \text{g} \quad (1)$$

$$m_{MAA} = d_{MAA} \times V_{MAA} = 1.015\ {\scriptstyle g/mL} \times 20\ \text{mL} = 20.3\ \text{g} \quad (2)$$

$$Yield(\%) = \frac{m_{expe}}{m_{MMA} + m_{MAA}} \times 100(\%) = \frac{32.1\ \text{g}}{18.8\ \text{g} + 20.3\ \text{g}} \times 100(\%) = 78.1(\%) \quad (3)$$

where m_{MMA} is the theoretical mass of the *MMA*, m_{MAA} is the theoretical mass of the *MAA*, d_{MMA} is the density of the *MMA*, d_{MAA} is the density of the *MAA*, V_{MMA} is the volume of the *MMA*, V_{MAA} is the volume of the *MAA*, and m_{expe} is the mass of the copolymer after the termination of the experiment.

We considered the initiator and crosslinker to have no contribution to the yield of the reaction. After drying, we obtained m_{expe} = 32.1 g of the *PMMA@P(MAA-co-EGDMA)* core-shell copolymer. Theoretically, 18.8 g of MMA and 20.3 g of *MAA* react, resulting in 39.1 g of product. The surplus amount of *MAA* (20.3 g − 18.8 g = 1.5 g) was added just to make sure that there would be enough *MAA* to form the *P(MAA-co-EGDMA)* shell. According to Equation (3), the *Yield*(%) of the polymerization reaction for the synthesis of the *PMMA@P(MAA-co-EGDMA)* core-shell copolymer was 78.1(%). Concerning the kinetics of the reaction, free radical polymerization started at a slow rate, but quickly accelerated, forming the polymer. Temperature plays a significant role in the polymerization reaction, as the initiator is activated at about 90 °C, producing the free radicals that polymerize the monomers [49–52].

Judging from the results of the different characterization techniques, it can be concluded that the product had a spherical shape at a nanoscale, with a very thin shell of *P(MAA-co-EGDMA)* around the core of the *PMMA*. FT-IR analysis showed peaks in the polymer's graph corresponding both to *PMMA* and *PMAA* bonds, while the SEM and TEM images confirmed its morphological properties. The morphological characterization also provided information about the composition of the product in regards to the two polymers, which was further demonstrated by the thermal analysis. The larger size of the core and the thinness of the shell exhibit advantages for potential applications, as the core could encapsulate larger amounts of other nanocompounds while the shell breaks smoothly to release the encapsulations. This is the case of delivery of antioxidants and other substances as well as adsorption and catalysis [53,54].

Recent literature on this subject by Peralta et al. [55] demonstrated the grafting of thermo-responsive polymers on inorganic compounds. However, these materials were activated at a temperature of 40 °C for drug delivery, while the copolymer in this study had a shell with a T_g of 228 °C, enabling its use at elevated temperatures for different applications such as incorporation in coatings. To that end, Wang et al. [56] synthesized lipophilic core-shell microspheres with inorganic shells. Nonetheless, the goal in the present research was to have both a polymeric core and a polymeric shell, and preferably for them to be hydrophilic. This aim was achieved using the PMAA as a compound of the shell. While Huang et al. have shown that the size of a core-shell copolymer can be tuned by following a specific method [37], that method requires an emulsion polymerization that is substantially more complex than the conventional radical polymerization proposed in the present work. Additionally, other researchers have synthesized core-shell compounds using specialized laboratory equipment and elaborate experimental paths [57–60] using catalysts [24] or by performing in situ processes [61].

Core-shell copolymers, both organic and inorganic, are very promising materials for several applications in the near future. Specifically, they can enhance the stability of crystal structures [62] and the responsiveness of sensors [63]. In addition, they increase the reaction activity of composite catalysts [64], have contributed to greener syntheses using natural plant products [65] and can also be very useful in magnetic applications [66]. It should be mentioned that core-shell polymers could play a significant role in effectively designing new multifunctional active materials for battery cells [67].

These applications notwithstanding, the simple synthesis discussed in this study was designed to be able to be performed in vitro, employing common experimental techniques and laboratory instruments without using intricate means for the production of the final product. Most importantly, the use of distilled water as the solvent/medium of the polymerization reaction is worth mentioning, due to the fact that most chemical routes nowadays continue to use expensive organic solvents to produce these or similar types of materials.

4. Conclusions

Following a simple experimental process of a two-step radical polymerization, a novel core-shell polymer was formed with a uniform size distribution, homogeneous spherical shape and thin shell around the polymeric core, without any admixtures. The novelty of this study is the relatively easy experimental setup and the fact that the copolymer exhibited different glass transition temperatures between the core and the shell, rendering the compound suitable for several applications. In addition, methacrylic acid was successfully polymerized as the shell in water, without forming a gel. This phenomenon was possible due to the already polymerized methyl methacrylate in the mixtures, which allowed the second monomer to be polymerized around the core. The produced core-shell copolymer was thoroughly characterized in respect of its morphology, structure, composition and thermal behavior. The results confirmed the successful synthesis of the material.

Author Contributions: Conceptualization, P.G. and I.A.K.; methodology, P.G. and I.A.K.; software, P.G. and I.A.K.; validation, P.G., I.A.K. and C.A.C.; formal analysis, P.G. and I.A.K.; investigation, P.G. and I.A.K.; resources, P.G. and I.A.K.; data curation, P.G.; writing—original draft preparation, P.G. and I.A.K.; writing—review and editing, I.A.K. and C.A.C.; visualization, P.G. and I.A.K.; supervision, C.A.C.; project administration, C.A.C.; funding acquisition, C.A.C. All authors have read and agreed to the published version of the manuscript.

Funding: This research was funded by the HORIZON 2020 Collaborative project "DECOAT" (Recycling of coated and painted textile and plastic materials, Grant agreement no.: 814505).

Conflicts of Interest: The authors declare no conflict of interest.

References

1. Ramli, R.A.; Laftah, W.A.; Hashim, S. Core–shell polymers: A review. *RSC Adv.* **2013**, *3*, 15543. [CrossRef]
2. Afshar, S.; Rashedi, S.; Nazockdast, H.; Ghazalian, M. Preparation and characterization of electrospun poly(lactic acid)-chitosan core-shell nanofibers with a new solvent system. *Int. J. Biol. Macromol.* **2019**, *138*, 1130–1137. [CrossRef] [PubMed]
3. Ali, S.S.M.; Eisa, W.H.; Abouelsayed, A. Solvent-free and large-scale preparation of silver@polypyrrole core@shell nanocomposites; structural properties and terahertz spectroscopic studies. *Compos. Part. B Eng.* **2019**, *176*, 107289. [CrossRef]
4. Balakumar, V.; Kim, H.; Manivannan, R.; Kim, H.; Ryu, J.W.; Heo, G.; Son, Y.A. Ultrasound-assisted method to improve the structure of CeO_2@polyprrole core-shell nanosphere and its photocatalytic reduction of hazardous Cr(6). *Ultrason. Sonochem.* **2019**, *59*, 104738. [CrossRef]
5. Anitha, B.; Nithiananthi, P. Oscillator strength and carrier dynamics in type I and inverted type I spherical core/ shell nanostructures under external laser field. *Superlattices Microstruct.* **2019**, *135*, 106288. [CrossRef]
6. Izu, N.; Uchida, T.; Itoh, T.; Shin, W. Decreasing the shell ratio of core-shell type nanoparticles with a ceria core and polymer shell by acid treatment. *Solid State Sci.* **2018**, *85*, 32–37. [CrossRef]
7. Hu, H.; Lin, Y.; Hu, Y.H. Synthesis, structures and applications of single component core-shell structured TiO_2: A review. *Chem. Eng. J.* **2019**, *375*, 122029. [CrossRef]
8. Han, J.; Wang, L.; Wang, L.; Li, C.; Mao, Y.; Wang, Y. Fabrication of a core-shell-shell magnetic polymeric microsphere with excellent performance for separation and purification of bromelain. *Food Chem.* **2019**, *283*, 1–10. [CrossRef]
9. Gul, S.; Shah, N.; Arain, M.B.; Rahman, N.; Rehan, T.; Ul-Islam, M.; Ullah, M.W.; Yang, G. Fabrication of magnetic core shell particles coated with phenylalanine imprinted polymer. *Polym. Test.* **2019**, *75*, 262–269. [CrossRef]
10. Farboudi, A.; Nouri, A.; Shirinzad, S.; Sojoudi, P.; Davaran, S.; Akrami, M.; Irani, M. Synthesis of magnetic gold coated poly (epsilon-caprolactonediol) based polyurethane/poly(N-isopropylacrylamide)-grafted-chitosan core-shell nanofibers for controlled release of paclitaxel and 5-FU. *Int. J. Biol. Macromol.* **2019**, *150*, 1130–1140. [CrossRef]
11. Karaxi, E.K.; Kartsonakis, I.A.; Charitidis, C.A. Assessment of Self-Healing Epoxy-Based Coatings Containing Microcapsules Applied on Hot Dipped Galvanized Steel. *Front. Mater.* **2019**, *6*, 222. [CrossRef]
12. Jabbar, T.A.; Ammar, S.H. Core/shell phosphomolybdic acid-supported magnetic silica nanocomposite (Ni@SiO_2-PMo): Synthesis, characterization and its application as a recyclable antibacterial agent. *Colloid Interface Sci. Commun.* **2019**, *33*, 100214. [CrossRef]
13. Amagat Molas, J.; Chen, M. Injectable PLCL/gelatin core-shell nanofibers support noninvasive 3D delivery of stem cells. *Int. J. Pharm.* **2019**, *568*, 118566. [CrossRef] [PubMed]

14. Zhou, Y.; Fang, A.; Wang, F.; Li, H.; Jin, Q.; Huang, L.; Fu, C.; Zeng, J.; Jin, Z.; Song, X. Core-shell lipid-polymer nanoparticles as a promising ocular drug delivery system to treat glaucoma. *Chin. Chem. Lett.* **2019**, *31*, 494–500. [CrossRef]
15. Wang, P.; Ji, Y.; Shao, Q.; Li, Y.; Huang, X. Core@shell structured Au@SnO_2 nanoparticles with improved N2 adsorption/activation and electrical conductivity for efficient N2 fixation. *Sci. Bull.* **2019**, *65*, 350–358. [CrossRef]
16. Shabzendedar, S.; Modarresi-Alam, A.R.; Noroozifar, M.; Kerman, K. Core-shell nanocomposite of superparamagnetic Fe_3O_4 nanoparticles with poly(m-aminobenzenesulfonic acid) for polymer solar cells. *Org. Electron.* **2019**, *77*, 105462. [CrossRef]
17. Pavlenko, M.; Siuzdak, K.; Coy, E.; Załęski, K.; Jancelewicz, M.; Iatsunskyi, I. Enhanced solar-driven water splitting of 1D core-shell Si/TiO_2/ZnO nanopillars. *Int. J. Hydrogen Energy* **2019**, *45*, 26426–26433. [CrossRef]
18. Wang, D.; An, Y.; Gao, S. Synthesis and characterization of urchin-like CuO nanorod/TiCu-based metallic glass core-shell powders with surface photovoltage performance. *Appl. Surf. Sci.* **2019**, *506*, 144871. [CrossRef]
19. Kawasaki, D.; Maeno, K.; Yamada, H.; Sueyoshi, K.; Hisamoto, H.; Endo, T. TiN-contained polymer-metal core-shell structured nanocone array: Engineering of sensor performance by controlling plasmonic properties. *Sens. Actuators B Chem.* **2019**, *299*, 126932. [CrossRef]
20. Hu, Y.; Wang, H.; Liu, D.; Lin, G.; Wan, J.; Jiang, H.; Lai, X.; Hao, S.; Liu, X. Lychee-like ZnO/$ZnFe_2O_4$ core-shell hollow microsphere for improving acetone gas sensing performance. *Ceram. Int.* **2019**, *46*, 5960–5967. [CrossRef]
21. Cheng, J.; Wang, J.; Yun, Y.; Rui, J.; Zhao, W.; Li, F. A novel core-shell structure reinforced Zr-based metallic glass composite with combined high strength and good tensile ductility. *J. Alloy. Compd.* **2019**, *803*, 413–416. [CrossRef]
22. Jia, E.; Zhao, S.; Shangguan, Y.; Zheng, Q. Toughening mechanism of polypropylene bends with polymer particles in core-shell structure: Equivalent rubber content effect related to core-shell interfacial strength. *Polymer* **2019**, *178*, 121602. [CrossRef]
23. Wang, J.; Xue, Z.; Li, Y.; Li, G.; Wang, Y.; Zhong, W.-H.; Yang, X. Synergistically effects of copolymer and core-shell particles for toughening epoxy. *Polymer* **2018**, *140*, 39–46. [CrossRef]
24. Wu, Y.; Zhang, Y.; Lv, X.; Mao, C.; Zhou, Y.; Wu, W.; Zhang, H.; Huang, Z. Synthesis of polymeric ionic liquids mircrospheres/Pd nanoparticles/CeO_2 core-shell structure catalyst for catalytic oxidation of benzyl alcohol. *J. Taiwan Inst. Chem. Eng.* **2019**, *107*, 161–170. [CrossRef]
25. Lynch, D.E.; Fellows, A.C.; Wilcock, R.; Sethi, S.; Armour, S.C.; Conteh, L. The use of poly(1-methylpyrrol-2-ylsquaraine) particles as a sacrificial template for the preparation of core-shell materials. *Mater. Chem. Phys.* **2019**, *227*, 163–169. [CrossRef]
26. Cao-Luu, N.-H.; Pham, Q.-T.; Yao, Z.-H.; Wang, F.-M.; Chern, C.-S. Synthesis and characterization of poly(N-isopropylacrylamide-co-N,N′-methylenebisacrylamide-co-acrylamide) core—Silica shell nanoparticles by using reactive surfactant polyoxyethylene alkylphenyl ether ammonium sulfate. *Eur. Polym. J.* **2019**, *120*, 109263. [CrossRef]
27. Gasaymeh, S.S.; Almansoori, N.N. Novel Formation Mechanism of Ag/PANI/PVP Core-Shell Nanocomposites. *Results Phys.* **2019**, *16*, 102882. [CrossRef]
28. Park, S. Complex core-shell morphologies of block copolymers revealed beneath the surface. *Appl. Surf. Sci.* **2019**, *494*, 309–314. [CrossRef]
29. Navin, K.; Kurchania, R. Structural, magnetic and electrochemical properties of LSMO-ZnO core-shell nanostructure. *Mater. Chem. Phys.* **2019**, *234*, 25–31. [CrossRef]
30. Khanal, A.; Inoue, Y.; Yada, M.; Nakashima, K. Synthesis of silica hollow nanoparticles templated by polymeric micelle with core-shell-corona structure. *J. Am. Chem Soc.* **2007**, *129*, 1534–1535. [CrossRef]
31. Wang, Y.; Rong, Z.; Tang, X.; Cao, S.; Chen, X.; Dang, W.; Wu, L. The design of scorodite@FeOOH core-shell materials and its stability treatment for arsenide. *Appl. Surf. Sci.* **2019**, *496*, 143719. [CrossRef]
32. Schroffenegger, M.; Reimhult, E. Thermoresponsive Core-Shell Nanoparticles and Their Potential Applications. *Compr. Nanosci. Nanotechnol.* **2019**, *2*, 145–170. [CrossRef]
33. Balaskas, A.C.; Kartsonakis, I.A.; Kordas, G.; Cabral, A.M.; Morais, P.J. Influence of the doping agent on the corrosion protection properties of polypyrrole grown on aluminium alloy 2024-T3. *Prog. Org. Coat.* **2011**, *71*, 181–187. [CrossRef]

34. Goulis, P.; Kartsonakis, I.; Konstantopoulos, G.; Charitidis, C. Synthesis and Processing of Melt Spun Materials from Esterified Lignin with Lactic Acid. *Appl. Sci.* **2019**, *9*, 5361. [CrossRef]
35. Goulis, P.; Kartsonakis, I.A.; Mpalias, K.; Charitidis, C. Combined effects of multi-walled carbon nanotubes and lignin on polymer fiber-reinforced epoxy composites. *Mater. Chem. Phys.* **2018**, *218*, 18–27. [CrossRef]
36. Goulis, P.; Konstantopoulos, G.; Kartsonakis, I.A.; Mpalias, K.; Anagnou, S.; Dragatogiannis, D.; Charitidis, C. Thermal Treatment of Melt-Spun Fibers Based on High Density PolyEthylene and Lignin. *C J. Carbon Res.* **2017**, *3*, 35. [CrossRef]
37. Huang, W.; Mao, Z.; Xu, Z.; Xiang, B.; Zhang, J. Synthesis and characterization of size-tunable core-shell structural polyacrylate-graft-poly(acrylonitrile-ran-styrene) (ASA) by pre-emulsion semi-continuous polymerization. *Eur. Polym. J.* **2019**, *120*, 109247. [CrossRef]
38. Weiss, A.V.; Koch, M.; Schneider, M. Combining cryo-TEM and energy-filtered TEM for imaging organic core-shell nanoparticles and defining the polymer distribution. *Int. J. Pharm.* **2019**, *570*, 118650. [CrossRef]
39. Dinc, M.; Esen, C.; Mizaikoff, B. Recent advances on core–shell magnetic molecularly imprinted polymers for biomacromolecules. *Trac. Trends Anal. Chem.* **2019**, *114*, 202–217. [CrossRef]
40. Ravindar Reddy, M.; Subrahmanyam, A.R.; Maheshwar Reddy, M.; Siva Kumar, J.; Kamalaker, V.; Jaipal Reddy, M. X-RD, SEM, FT-IR, DSC Studies of Polymer Blend Films of PMMA and PEO. *Mater. Today Proc.* **2016**, *3*, 3713–3718. [CrossRef]
41. Mekuria, T.D.; Zhang, C.; Fouad, D.E. The effect of thermally developed SiC@SiO_2 core-shell structured nanoparticles on the mechanical, thermal and UV-shielding properties of polyimide composites. *Compos. Part. B Eng.* **2019**, *173*, 106917. [CrossRef]
42. Tommasini, F.J.; Ferreira, L.d.C.; Tienne, L.G.P.; Aguiar, V.d.O.; Silva, M.H.P.d.; Rocha, L.F.d.M.; Marques, M.d.F.V. Poly (Methyl Methacrylate)-SiC Nanocomposites Prepared Through in Situ Polymerization. *Mater. Res.* **2018**, *21*, 1–7. [CrossRef]
43. Yuksel, N.; Baykara, M.; Shirinzade, H.; Suzen, S. Investigation of triacetin effect on indomethacin release from poly(methyl methacrylate) microspheres: Evaluation of interactions using FT-IR and NMR spectroscopies. *Int J. Pharm.* **2011**, *404*, 102–109. [CrossRef] [PubMed]
44. Sivy, G.T.; Coleman, M.M. Fourier transform ir studies of the degradation of polyacrylonitrile copolymers—II. *Carbon* **1981**, *19*, 127–131. [CrossRef]
45. Kartsonakis, I.A.; Danilidis, I.L.; Pappas, G.S.; Kordas, G.C. Encapsulation and release of corrosion inhibitors into titania nanocontainers. *J. Nanosci. Nanotechnol.* **2010**, *10*, 5912–5920. [CrossRef]
46. Eissa, A.S. Effect of SDS on whey protein polymers. Molecular investigation via dilute solution viscometry and dynamic light scattering. *Food Hydrocoll.* **2019**, *87*, 97–100. [CrossRef]
47. Makan, A.C.; Spallek, M.J.; du Toit, M.; Klein, T.; Pasch, H. Advanced analysis of polymer emulsions: Particle size and particle size distribution by field-flow fractionation and dynamic light scattering. *J. Chromatogr. A* **2016**, *1442*, 94–106. [CrossRef]
48. Cametti, C.; D'Amato, R.; Furlani, A.; Russo, M.V. Dynamic light scattering and optical absorption study of poly(monosubstituted)acetylene polymers and copolymers. *Chem. Phys. Lett.* **2003**, *370*, 602–608. [CrossRef]
49. Gao, H.; Waechter, A.; Konstantinov, I.A.; Arturo, S.G.; Broadbelt, L.J. Application and comparison of derivative-free optimization algorithms to control and optimize free radical polymerization simulated using the kinetic Monte Carlo method. *Comput. Chem. Eng.* **2018**, *108*, 268–275. [CrossRef]
50. Sankar, K.; Rajendran, V. Ultrasound assisted free radical polymerization of glycidyl methacrylate by a new disite phase-transfer catalyst system: A kinetic study. *Ultrason. Sonochem.* **2012**, *19*, 1205–1212. [CrossRef]
51. Prabha, J.; Susan Jemima, W.; Jayaprada, M.; Umapathy, M.J. Synergistic effect of ultrasonication and phase transfer catalysts in radical polymerization of methyl methacrylate—A kinetic study. *Ultrason. Sonochem.* **2017**, *35*, 333–341. [CrossRef] [PubMed]
52. Marimuthu, E.; Murugesan, V. Influence of ultrasonic condition on phase transfer catalyzed radical polymerization of methyl methacrylate in two phase system—A kinetic study. *Ultrason. Sonochem.* **2017**, *38*, 560–569. [CrossRef] [PubMed]
53. Van Tran, V.; Loi Nguyen, T.; Moon, J.-Y.; Lee, Y.-C. Core-shell materials, lipid particles and nanoemulsions, for delivery of active anti-oxidants in cosmetics applications: Challenges and development strategies. *Chem. Eng. J.* **2019**, *368*, 88–114. [CrossRef]
54. Su, H.; Tian, Q.; Hurd Price, C.-A.; Xu, L.; Qian, K.; Liu, J. Nanoporous core@shell particles: Design, preparation, applications in bioadsorption and biocatalysis. *Nano Today* **2020**, *31*, 100834. [CrossRef]

55. Peralta, M.E.; Jadhav, S.A.; Magnacca, G.; Scalarone, D.; Martire, D.O.; Parolo, M.E.; Carlos, L. Synthesis and in vitro testing of thermoresponsive polymer-grafted core-shell magnetic mesoporous silica nanoparticles for efficient controlled and targeted drug delivery. *J. Colloid Interface Sci.* **2019**, *544*, 198–205. [CrossRef] [PubMed]
56. Wang, T.; Bao, Y.; Gao, Z.; Wu, Y.; Wu, L. Synthesis of mesoporous silica-shell/oil-core microspheres for common waterborne polymer coatings with robust superhydrophobicity. *Prog. Org. Coat.* **2019**, *132*, 275–282. [CrossRef]
57. Zhang, W.; Kong, Y.; Jin, X.; Yan, B.; Diao, G.; Piao, Y. Supramolecule-assisted synthesis of cyclodextrin polymer functionalized polyaniline/carbon nanotube with core-shell nanostructure as high-performance supercapacitor material. *Electrochim. Acta* **2020**, *331*, 135345. [CrossRef]
58. Yang, X.; Wan, G.; Ma, S.; Xia, H.; Wang, J.; Liu, J.; Liu, Y.; Chen, G.; Bai, Q. Synthesis and optimization of SiO_2@SiO_2 core-shell microspheres by an improved polymerization-induced colloid aggregation method for fast separation of small solutes and proteins. *Talanta* **2020**, *207*, 120310. [CrossRef]
59. Pourjavadi, A.; Rahemipoor, S.; Kohestanian, M. Synthesis and characterization of multi stimuli-responsive block copolymer-silica hybrid nanocomposite with core-shell structure via RAFT polymerization. *Compos. Sci. Technol.* **2020**, *188*, 107951. [CrossRef]
60. Shevchenko, N.; Pankova, G.; Laishevkina, S.; Iakobson, O.; Koshkin, A.; Shabsels, B. Core-shell polymer particles containing derivatives of 1,3-diphenyl-β-diketonate boron difluoride: Synthesis and spectroscopic investigation of toluene vapor sorption. *Colloids Surf. A Physicochem. Eng. Asp.* **2019**, *562*, 310–320. [CrossRef]
61. Zhang, S.; Wen, Y.; Wang, L.; Yang, Y.; Liu, S. In situ synthesis of a multifunctional polymer with a stable core-shell structure for effective dewatering. *Miner. Eng.* **2019**, *141*, 105858. [CrossRef]
62. Zhang, H.; Jiao, Q.; Zhao, W.; Guo, X.; Li, D.; Sun, X. Enhanced Crystal Stabilities of ε-CL-20 via Core-Shell Structured Energetic Composites. *Appl. Sci.* **2020**, *10*, 2663. [CrossRef]
63. Bui, Q.C.; Largeau, L.; Morassi, M.; Jegenyes, N.; Mauguin, O.; Travers, L.; Lafosse, X.; Dupuis, C.; Harmand, J.-C.; Tchernycheva, M.; et al. GaN/Ga_2O_3 Core/Shell Nanowires Growth: Towards High Response Gas Sensors. *Appl. Sci.* **2019**, *9*, 3528. [CrossRef]
64. Li, Y.; Wang, Q.; Wang, D.; Yan, X. NO-CH4-SCR Over Core-Shell MnH-Zeolite Composites. *Appl. Sci.* **2019**, *9*, 1773. [CrossRef]
65. Khatami, M.; Alijani, H.; Nejad, M.; Varma, R. Core@shell Nanoparticles: Greener Synthesis Using Natural Plant Products. *Appl. Sci.* **2018**, *8*, 411. [CrossRef]
66. Obaidat, I.; Nayek, C.; Manna, K. Investigating the Role of Shell Thickness and Field Cooling on Saturation Magnetization and Its Temperature Dependence in Fe_3O_4/γ-Fe_2O_3 Core/Shell Nanoparticles. *Appl. Sci.* **2017**, *7*, 1269. [CrossRef]
67. Dell'Era, A.; Scaramuzzo, F.A.; Stoller, M.; Lupi, C.; Rossi, M.; Passeri, D.; Pasquali, M. Spinning Disk Reactor Technique for the Synthesis of Nanometric Sulfur TiO_2 Core–Shell Powder for Lithium Batteries. *Appl. Sci.* **2019**, *9*, 1913. [CrossRef]

Publisher's Note: MDPI stays neutral with regard to jurisdictional claims in published maps and institutional affiliations.

fibers

MDPI

Article

Low-Cost Electrodeposition of Size-Tunable Single-Crystal ZnO Nanorods

Elias Sakellis [1,2], Antonis Markopoulos [1], Christos Tzouvelekis [1], Manolis Chatzigeorgiou [1], Anastasios Travlos [1] and Nikos Boukos [1,*]

[1] National Centre for Scientific Research "Demokritos", Institute of Nanoscience and Nanotechnology, 15310 Agia Paraskevi Attikis, Greece; e.sakellis@inn.demokritos.gr (E.S.); antmarko@hotmail.com (A.M.); christzouvelekis@gmail.com (C.T.); e.chatzigeorgiou@inn.demokritos.gr (M.C.); a.travlos@inn.demokritos.gr (A.T.)

[2] Section of Solid State Physics, Physics Department, University of Athens, Panepistimiopolis, Zografos, 15684 Athens, Greece

* Correspondence: n.boukos@inn.demokritos.gr

Abstract: In this paper we report a low cost, simple, electrochemical method for large-area growth of single crystal ZnO nanorods. The method utilizes a metallic zinc foil as the source of the necessary zinc ions for ZnO growth on indium-doped tin oxide (ITO) glass slides. The method is thoroughly discussed and investigated varying all the parameters involved. The resulting ZnO nanorods are highly oriented along c-axis and densely packed, while their length and diameter can be tuned by varying the growth parameters. Two different types of seed layers on the ITO glass slides are tested. A seed layer made by spin coating of ZnO nanoparticles results in a twofold increase of the ZnO nanorod surface density as compared with a ZnO thin film seed layer by physical vapor deposition. Additionally, the effect of oxygen supply during electrodeposition was investigated as a crucial regulatory parameter not only for the geometrical and topological characteristics of the ZnO nano-arrays but for their physical properties as well.

Keywords: ZnO; nanorods; electrochemical method; oxygen; nanoparticles; electrodeposition

Citation: Sakellis, E.; Markopoulos, A.; Tzouvelekis, C.; Chatzigeorgiou, M.; Travlos, A.; Boukos, N. Low-Cost Electrodeposition of Size-Tunable Single-Crystal ZnO Nanorods. *Fibers* **2021**, *9*, 38. https://doi.org/10.3390/fib9060038

Academic Editor: Vijay Kumar Thakur

Received: 16 April 2021
Accepted: 1 June 2021
Published: 7 June 2021

1. Introduction

Among the transparent oxides, ZnO stands of great importance. It is a wide and direct bandgap (E_g = 3.23–3.42 eV), II–VI semiconductor, with relatively large exciton binding energy (about 60 meV) at room temperature and high electron mobility [1]. The aforementioned -among others- properties are the reason of a plethora of applications concerning ZnO. Solar cells, photocatalytic cells, sensors and optoelectronics are only few of these fields. [2–6] There are several techniques used for zinc oxide growth with prime desiderata the high quality and the low cost [7]. Electrodeposition is a promising candidate method which attempts to fulfil the previously cited needs. The main advantage of electrodeposition in general is its simplicity, while, varying at will the electrochemical parameters affects the structural as well as the physical properties of the resultant ZnO layers [8–12]. To the best of our knowledge, the established electrodeposition methods utilize zinc salts as the source of the required zinc ions for the growth of ZnO nanostructures, as well as other salts for supporting electrolytes [9,13–16]. The use of chloride or nitrate salts may result in intentional [14,15] or unintentional [16] incorporation of anions impurities such as Cl^- in the ZnO structure affecting ZnO growth and electronic properties.

In this work a very simple, fast, low temperature, low equipment cost and time expense, electrodeposition method for growing ZnO nanorods is presented. Its simplicity originates from the use of only two electrodes as compared to the commonly reported three electrode electrodeposition methods, while it does not require salts for supporting electrolyte or any pH adjustment [9,13–16]. The combination of a zinc metallic foil used as counter-electrode and aqueous formamide solution employed as electrolyte provides the necessary zinc for the growth of ZnO, whereas no additional metal cations or chloride/nitrate anions are present that may lead to the incorporation of undesirable impurities in the ZnO crystal [14–16]. The proposed electrodeposition method is about ten times faster than previously reported chemical bath ones utilizing formamide solutions [17], being comparable with the established three electrode methods employing zinc salts [9]. The resulting ZnO rods are homogeneously grown, highly oriented and densely packed while their diameter is controllable. The method has been tested on indium-doped tin oxide (ITO) coated glass substrates utilizing two different seed layers: (i) a thin ZnO film prepared by Electron-Beam Physical Vapor Deposition (EBPVD), (procedure A) and (ii) a thin film made by spin-coating of ZnO nanoparticles (procedure B). Furthermore, the morphology and properties of the ZnO nanorods was investigated changing the oxygen concentration in the electrodeposition bath solution, while using a spin-coating seed layer (procedure C).

2. Materials and Methods

2.1. Materials and Cleaning Procedure

Indium-doped tin oxide ITO sputtered glasses (5–15 Ω/□ at room temperature), purchased from Delta Technologies Ltd., Loveland, CO, USA, were used as substrates after thorough cleaning. This step was crucial for the subsequent ZnO layers adherence to the substrate and homogeneity. At first, the glasses were sonicated within a 2% v/v solution of alkaline liquid concentrate (Hellmanex III) in de-ionized water (DI water) for 30 min and after that meticulously rinsed with DI water. Next, the glasses were ultrasonically cleaned in acetone and IPA, for 10 min each, rinsing with DI water in-between. Finally, the substrates were cleaned with a solution of 3% HCl acid and 1% HNO_3 acid, rinsed with DI water and dried under nitrogen gas flow.

Zinc foils (99.9% Sigma-Aldrich, St. Louis, MO, USA) were used after been carefully cleaned with acetone and absolute ethanol.

Finally, analytical grade formamide (Sigma Aldrich, St. Louis, MO, USA) was used to prepare aqueous solutions with DI water.

2.2. Preparation of ZnO Seed Layers

2.2.1. EBPVD

ZnO pellets, initially made by cold pressing of ZnO powder (99.9995% Alfa Aesar, Ward Hill, MA, USA), were used as target material in a high vacuum chamber (10^{-4} Pa base pressure), equipped with 2 electron guns. ZnO thin films of 50 nm thickness were deposited by e-beam evaporation on the ITO/glass substrates at room temperature. The deposition rate varied from 1.2 nm/s to 1.5 nm/s, as monitored by a calibrated quartz crystal microbalance. Next, the seeded substrates were annealed at 300 °C for 15 min in air.

2.2.2. ZnO Nanoparticles (NPs)

ZnO nanoparticles were synthesized according to Haase et al. [18] and with the combination of other, similar to the previous [19], methods. In particular, 2.95 gr zinc acetate dehydrated, and 1.48 gr KOH (0.5 M) were dissolved in 125 mL and 65 mL of methanol, respectively. The first solution was heated to 60 °C under vigorous stirring and then the potassium hydroxide solution was gradually added within 15 min. The resulting solution was vigorously stirred for 2 h at 60 °C. Next, the solution was left to sit without heating or stirring for another two hours. Subsequently, the supernatant liquid was removed from the precipitated nanoparticles, 50 mL of methanol was added, and the solution was centrifuged. Finally, 5 mL of chloroform and 25 mL of methanol was added to the residuum nanoparticles. The final solution of the nanoparticles was used to spin-coat the ITO glasses with a thin film of ZnO nanoparticles. Lastly, the substrates were annealed for 15 min at 300 °C in air to improve cohesion of the ZnO thin film on the ITO glasses.

2.3. Electrochemical Growth of ZnO Nanorods (NRs)

The electrolyte was a formamide aqueous solution of various concentrations (Table 1) containing Zn complexes, which was created by immersing a cleaned zinc foil in an aqueous solution formamide for 18 h. The ZnO seeded substrate was used as the working electrode while a cleaned zinc foil was used as the counter electrode. The 1.25 cm × 3.50 cm electrodes were positioned at a distance of 2.5 cm, facing each other. A constant voltage of 0.3 V, with respect to the Zn foil, was applied to the working electrode by a Metrohm Autolab PGSTAT204 potentiostat. The electrolyte temperature was kept constant at 65 °C ± 1 °C in a bath. The solution was steered, and the current was monitored throughout the electrodeposition. Growth times could be varied from a few minutes to few hours, depending on the desirable dimensions of the ZnO formed. After growth, the samples were rinsed in distilled water and ethanol, and finally dried by nitrogen gas flow.

Table 1. Experimental data for different formamide concentrations for each of the three procedures A, B and C: formamide concentration C, length L_{rod}, diameter D_{rod} and the aspect ratio L_{rod}/D_{rod} of ZnO nanorods, the transferred charge during electrochemical growth, the density of ZnO nanorods, the optical energy gap $E_{g,opt}$ calculated from UV-vis measurements and the wavelength of the maximum intensity from PL spectra.

Formamide C (% vol)	L_{rod} (nm)	D_{rod} (nm)	L_{rod}/D_{rod}	Charge (C)	Density ρ (rods/μm²)	$E_{g,opt}$ (eV)	PL Peak (nm)
Procedure A — Growth Method: Electrodeposition Seeding Layer: ZnO Film (EBPVD) O_2 Supply: No							
0.05%	480	70	6.9	1.206	31	3.31	396
0.1%	640	75	8.5	1.378	38	3.34	380
0.5%	750	85	8.8	2.374	23	3.31	376
1%	900	95	9.5	5.452	21	3.33	381
Procedure B — Growth Method: Electrodeposition Seeding Layer: Chem. Synthesized ZnO NPs (Spin Coating) O_2 Supply: No							
Formamide C (% vol)	**L_{rod} (nm)**	**D_{rod} (nm)**	**L_{rod}/D_{rod}**	**Charge (C)**	**Density ρ (rods/μm²)**	**$E_{g,opt}$ (eV)**	**PL Peak (nm)**
0.05%	300	40	7.5	1.246	60	3.31	380
0.1%	490	50	9.8	1.527	64	3.31	380
0.5%	690	65	10.6	3.226	45	3.33	380
1%	810	75	10.8	4.099	43	3.35	376
Procedure C — Growth Method: Electrodeposition Seeding Layer: Chem. Synthesized ZnO NPs (Spin Coating) O_2 Supply: Yes							
Formamide C (% vol)	**L_{rod} (nm)**	**D_{rod} (nm)**	**L_{rod}/D_{red}**	**Charge (C)**	**Density ρ (rods/μm²)**	**$E_{g,opt}$ (eV)**	**PL Peak (nm)**
0.05%	590	80	7.4	2.217	30	3.27	381
0.1%	630	75	8.4	2.883	32	3.26	380
0.5%	1040	100	10.4	6.223	30	3.30	381
1%	1500	125	12.0	9.787	25	3.31	380

2.4. Characterization

The morphology of the resulting nanostructures was determined by scanning electron microscopy (SEM) using an FEI Inspect microscope, operating at a voltage of 25 kV. Transmission Electron Microscopy (TEM) was undertaken utilizing an FEI CM20 microscope operating at 200 kV. TEM specimens of the grown nanostructures were prepared by scraping off the substrates and placing them on carbon support copper TEM grids. In order to examine the ZnO nanoparticles 5 μL of the final solution were drop-casted on the same type of TEM grids followed by evaporation at ambient conditions. Powder X-ray diffraction (XRD) spectra, with 2θ range 25–80°, were acquired with a SIEMENS D500 Bragg Brentano diffractometer, using Cu Kα radiation (λ = 0.15418 nm), step-size 0.03° and measuring time 2 s/step. Photoluminescence (PL) measurements were carried out using a Horiba Jobin-Yvon iHR320 Spectrometer with a He–Cd laser (325 nm) as the excitation source. UV-Visible (UV-Vis) absorption measurements were obtained with a Shimadzu, UV2100 UV-VIS split-beam spectrophotometer in a wavelength range of 250 nm to 900 nm, employing an ITO glass-slide as reference.

3. Results and Discussion

Several experiments have been carried out in order to investigate thoroughly the role of each one of the parameters involved in the growth process. As mentioned in the introduction, three procedures were carried out that aim not only to present and investigate this growth method but also to examine the role of the seed layer on the growth (comparing procedures A and B), as well as to control the dimensions and photoluminescence of the grown ZnO nanorods (comparing procedures B and C). It should be mentioned that all subsequent results concern a typical two hour electrodeposition growth.

The ZnO growth mechanism could be described by the following steps: Zinc-formamide complexes are formed in the vicinity of the Zn foil, counter-electrode and are subsequently transferred to the substrate, working electrode, under the applied electrical field. There, Zn reacts with oxygen and finally ZnO is formed. The two reactions which take place during growth are:

$$Zn + nHCONH_2 + H_2O \rightarrow [Zn(HCONH_2)_n]^{2+} + 2e^- \quad (1)$$

$$[Zn(HCONH_2)_n]^{2+} + \frac{1}{2}O_2 + 2e^- \rightarrow ZnO + nHCONH_2 \quad (2)$$

at the Zn foil and at the substrate region, respectively [17,20,21].

Scanning electron microscopy images (Figure 1), transmission electron microscopy images (Figure 2), X-ray diffraction spectra (Figure 3), photoluminescence measurements (Figure 4) and UV-Visible absorption measurements (Figure 5), provide information for the topology, crystallinity, the composition and the electronic structure of ZnO.

From the SEM images (Figure 1), regardless of which procedure was followed, it becomes apparent that the substrate is uniformly covered with hexagonal ZnO rods preferentially oriented perpendicular to the substrate. It should be noted that the synthesized nanorod arrays display lateral homogeneity across the whole substrate area, i.e., about 33×10 mm^2, making this growth method promising to be upscaled to larger areas. X-ray diffraction patterns (Figure 4) verifies that that no other phase except for wurtzite ZnO (JPDS 36-1451) was grown during this electrochemical method. Furthermore, depicted are ITO peaks (JPDS 6-416) originating from the substrate. The texture coefficient TC for the ZnO (002) peak is calculated [22] from the following equation:

$$TC_{(002)} = \frac{I_{(002)}/I^0_{(002)}}{1/N \sum I_{(hkl)}/I^0_{(hkl)}} \quad (3)$$

where N is the number of the diffraction peaks, $I_{(hkl)}$ and $I^0_{(hkl)}$ are the measured and reported in the JCPDS 36-1451 intensities, respectively. The $TC_{(002)}$ values are 3.5, 3.8 and 4.8 for procedures A, B and C, respectively. These high values indicate that the nanorods are textured and well-aligned with the c-axis perpendicular to the substrate [22,23], in agreement with the SEM results. Moreover, the degree of texturing increases from Procedure A to Procedure C. It has been reported that the growth of the more initially inclined rods is impeded, whereas it is favored for the more vertically oriented ones for geometry reasons [22,23]. As can be seen in Table 1, the surface density of ZnO nanorods grown by Procedure B is twice that by Procedure A, leaving less available space for the growth of inclined nanorods, thus increasing texturing coefficient. On the other hand, nanorods grown by Procedure C have the highest length and aspect ratio, leading to the maximum value of TC.

Dimensional information deduced from the SEM images is quantified (Table 1) and plotted (Figure 6). The general trend is that the length, diameter and aspect ratio increase when formamide concentration is increased in each procedure, since there are more Zn atoms available, as transferred charge confirms (see Table 1).

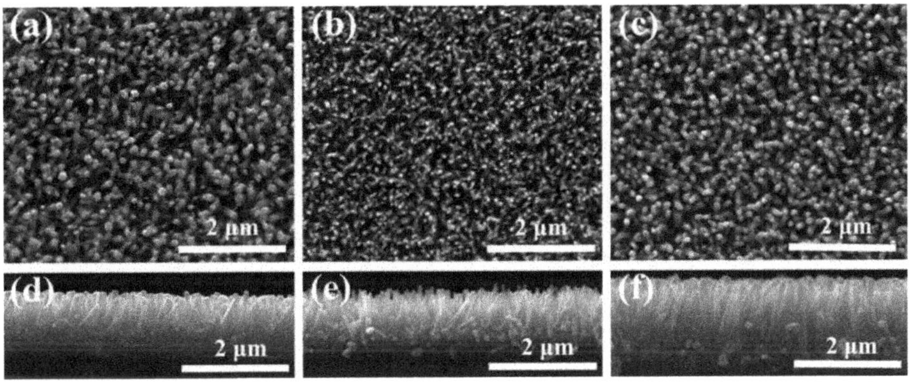

Figure 1. SEM images—plane view and cross section- of the ZnO nanorods grown in 0.5% formamide for each of the three procedures A (**a**,**d**), B (**b**,**e**) and C (**c**,**f**) respectively.

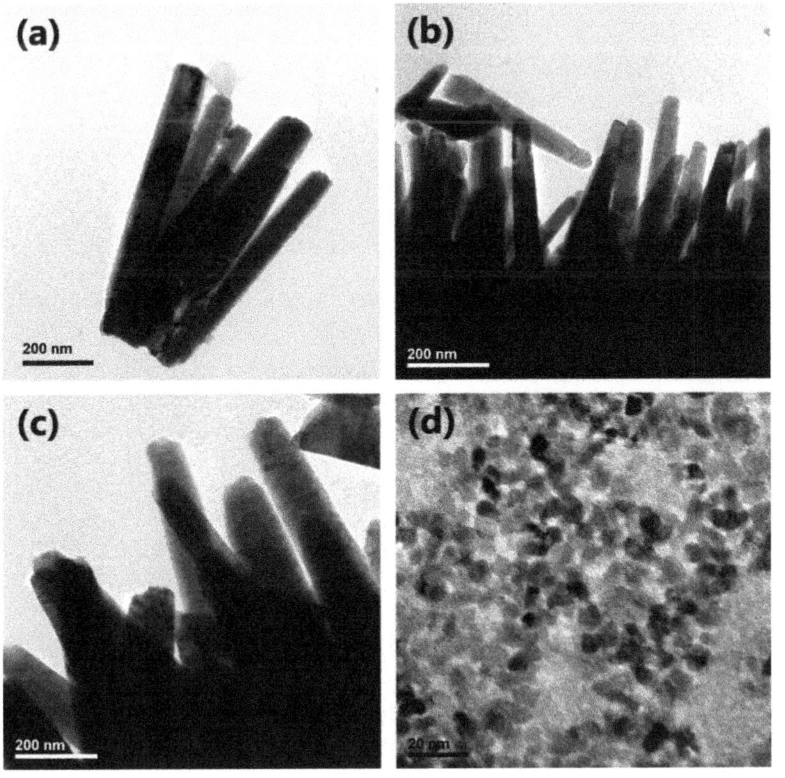

Figure 2. Bright field TEM images of the ZnO nanorods grown in 0.5% formamide for each of the three procedures A (**a**), B (**b**) and C (**c**) and of ZnO nanoparticles used for the spin-coating seed layer (**d**).

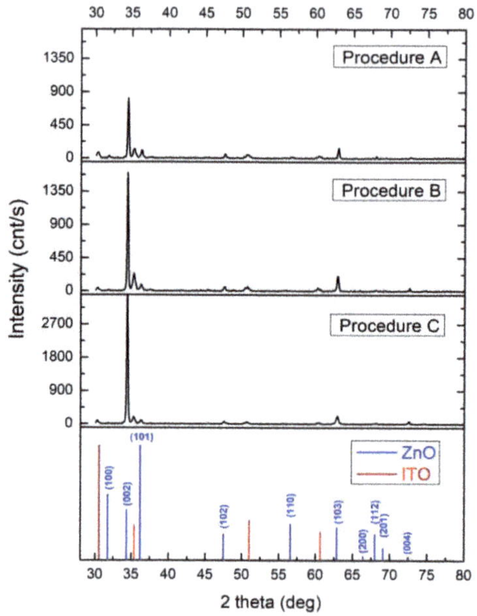

Figure 3. X-ray diffraction spectrum of ZnO nanorods grown in 0.5% formamide for each of the three procedures A, B and C.

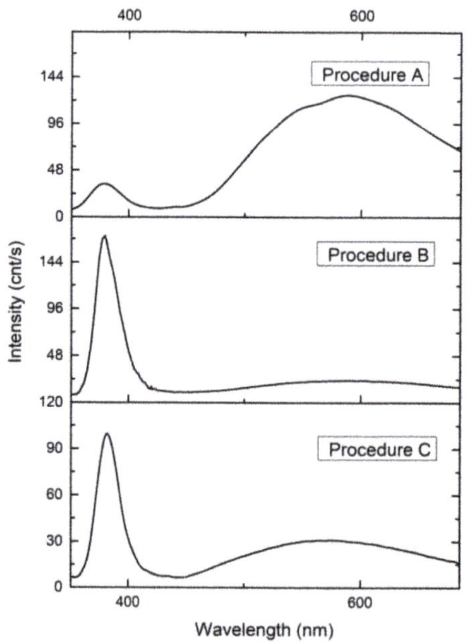

Figure 4. Room temperature photoluminescence spectra of the ZnO nanorods grown in 0.5% formamide for each of the three procedures A, B and C.

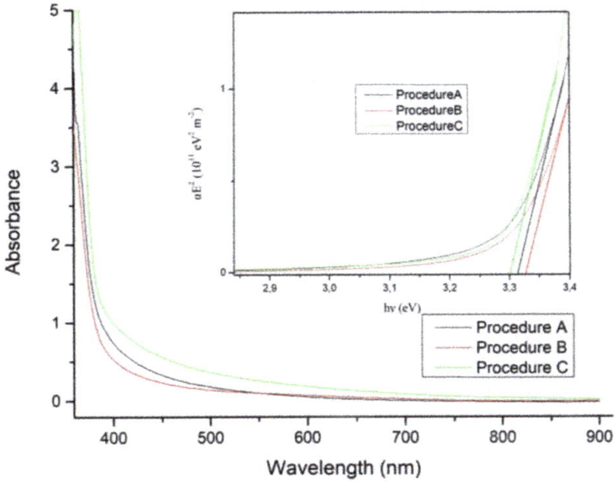

Figure 5. UV-vis transmittance spectra of the ZnO nanorods grown in 0.5% formamide for each of the three procedures A, B and C. Inset: $(\alpha E)^2$ versus E—where E is the photon energy and α is the absorption coefficient.

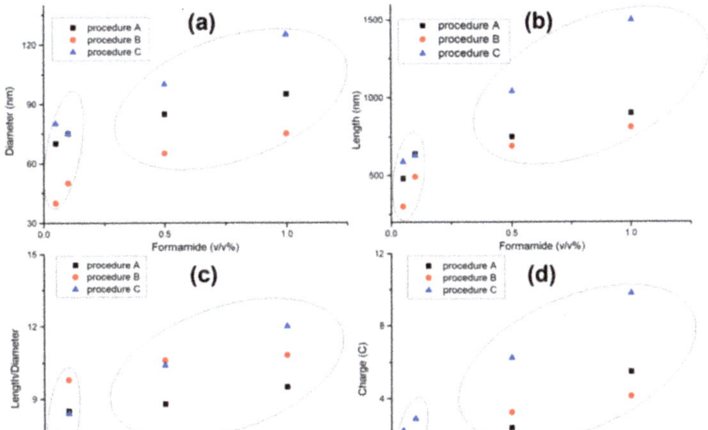

Figure 6. Diameter (**a**), length (**b**), and aspect ratio (**c**) of the ZnO nanorods as well as the charge (**d**) transferred during electrodeposition for each of the three procedures A, B and C.

On the other hand, the density of ZnO rods appears to increase at first and then as formamide concentration is further increased, the density of rods decreases in each procedure. Raising formamide concentration leads to an increase of Zn complexes that results to an increment of "starting points" on the seed layer and therefore, to a raise in rod density. As formamide concentration is further increased, the density of "starting points" on the seed layer exceeds some value above which nanorods are growing very close to each other resulting to their coalescence and consequent reduction of their density.

It is well known that there is an anisotropic growth along c-axis, since the growth rate in <001> direction is larger than that of <101> and <100> [24–26]. Thus, different growth rates are expected in the two directions, the one parallel r_p and the other vertical r_v to the growth *axis*. The ratio of these rates determines the aspect ratio of each rod, i.e., $r_p/r_v \sim 1/d$. Moreover, if the ratio $r_p/r_v(t)$ varies as a function of time the diameter d of the rod will change along its length l. Since the diameter of each nanorod diminishes along its growth direction (Figures 1 and 3), it can be concluded that $r_p/r_v(t)$ is an ascending function of time. This could be attributed to a gradual change of the electrochemical environment—for example the local electric field—in the vicinity of the corresponding surfaces of the growing nanorods resulting in a pyramidal shape of the nanorods (Figure 3).

The role of the seeding layer on the electrochemical growth of ZnO nanorods can be determined comparing the nanorods grown utilizing procedures A and B. It is evident from the data presented in Table 1, that a major difference is the surface density of the nanorods grown by the two procedures, i.e., a twofold increase of the nanorods density is achieved when a spin-coating seed layer (procedure B) is used instead of an EBPVD thin film layer (procedure A). This difference can be explained as follows: ZnO nanocrystals on the seeding layer act as preferred nucleation centers for the growth of ZnO nanorods. The mean crystallite size of the seed layer in procedure B is 3–8 nm corresponding to the mean ZnO nanoparticle size shown in the TEM micrograph (Figure 2d). On the other hand, the mean crystallite size of the EBPVD thin film layer used in procedure A is expected to be in the range of its thickness [27], that is roughly 20–40 nm. Consequently, the smaller crystallite size in the spin coating seed layer leads to a higher nuclei density for the initial ZnO precipitation, resulting in an increased surface density of the ZnO nanorods grown by procedure B.

UV-Visible (UV-Vis) absorption measurements (Figure 6) allowed for an estimation of the optical energy gap $E_{g,opt}$ (Table 1), by plotting (Figure 6 inset) the $(\alpha E)^2$ versus E—where E is the photon energy and α is the absorption coefficient—and extrapolating the (increasing) linear part of the curve. The optical energy gap was evaluated according to Equation (4) that describes the energy dependence E of the absorption coefficient α [28] in the case of direct band gap semiconductors:

$$\alpha \propto \sqrt{h\nu - E_{g,opt}}/(\mathrm{h}\nu) \tag{4}$$

The optical energy gaps calculated from UV-Vis measurements for each one of the procedures and for various formamide concentrations are tabulated in Table 1.

The PL spectra of zinc oxide nanorods for each procedure are presented at Figure 5. All the PL spectra exhibit a peak at 380 nm which is attributed to the recombination of free excitons [29]. Moreover, a broad peak including green (around 520 nm) and yellow-orange (around 600 nm) is observed [30]. This broad band is attributed to intrinsic defects of ZnO [10,31,32]. More specifically, this band is due to several kinds of defects· the green between 490 nm and 530 nm, is frequently attributed to oxygen vacancies, zinc vacancies, zinc interstitials, oxygen antisites and transitions between the latter two [33–43]. The yellow luminescence (around 590 nm) is usually attributed to interstitial oxygen [43,44]. Thus, supplying additional oxygen during growth (procedure C) results in an enhancement of the aforementioned part of the spectrum as compared to that corresponding to procedure B during which no excess oxygen is supplied.

The present growth method is versatile and has several advantages over previously proposed growth methods. It is reported that ZnO nanorods can be grown utilizing a Zn foil and a seeded substrate immersed in a formamide solution at about 65 °C for several hours (24 h [17], 60 h [20], 20 h [24]). With the present method, the procedure is greatly accelerated, while the good quality of the crystal is preserved. The electric field facilitates the constant flow of zinc-formamide complexes resulting in an enhanced ZnO growth rate at least an order of magnitude higher than the ones previously reported [9,17,24]. Furthermore, whereas most of the reported methods [9,13–16] use salts in the growth solution, that act either as zinc source or support conductivity of the electrolyte, the present method employs only a metallic zinc foil. Hence there are no impurities that could be incorporated in the ZnO crystal during growth. Changing formamide concentration alters the availability of

Zn complexes. Whilst the ratio of zinc to oxygen concentration is lower than one, raising the formamide concentration up to ~0.1% results to an approximately linear increment of the growth rate. As the formamide concentration is further increased, the ratio of zinc to oxygen concentration tends to one and the growth rate decreases with a saturation trend (Figure 2, Procedures A and B). Additional oxygen supply lowers the aforementioned ratio of concentrations, and the linear growth rate is extended to higher zinc concentrations (Figure 2, Procedure C). Thus, control of oxygen inflow and zinc complexes supply, can be used to tailor the ZnO nanorods dimensions as well as to incorporate excess oxygen or zinc atoms in the crystal structure, altering the type of the intrinsic defects and resulting to different physical properties of the ZnO nanorods.

4. Conclusions

Single crystalline ZnO nanorods of controllable length, diameter and density were grown by a novel, simple and electrodeposition method on seeded ITO glass substrates. The method utilizes a counter electrode of metallic zinc foil to provide the necessary zinc ions, while no reference electrode, no salts (employed as zinc source or supporting electrolyte) and no pH adjustment are required. The absence of cations other than those of Zn ensures that no impurities are incorporated in the ZnO structure during growth. The ZnO nanorod arrays are vertically well-aligned and textured with the *c-axis* perpendicular to the substrate, while they are homogeneously distributed over the whole substrate area of about 3 cm^2, making this growth method promising to be upscaled to larger areas. The increase of concentration of the organic solvent results in an enhancement of the growth rate and aspect ratio of the ZnO nanorods due to the increased supply of Zn ions, while oxygen bubbling during growth, apart from enhancing growth rate, also increases the yellow luminescence that is attributed to the formation of oxygen interstitials in the ZnO lattice. Finally, the utilization of a spin-coated seeding layer, comprised of ZnO nanoparticles, doubles the ZnO nanorod surface density compared to that of an EBPVD seeding layer, due to the smaller crystallite size of the former.

Author Contributions: Conceptualization, E.S. and N.B.; methodology, A.M., A.T. and E.S.; investigation, formal analysis, A.M., C.T. and M.C.; writing—original draft preparation, E.S. and N.B.; writing—review, A.T. All authors have read and agreed to the published version of the manuscript.

Funding: This research was partly funded by the project "National Infrastructure in Nanotechnology, Advanced Materials and Micro-/Nanoelectronics" (MIS 5002772) which is implemented under the Action "Reinforcement of the Research and Innovation Infrastructure", funded by the Operational Programme "Competitiveness, Entrepreneurship and Innovation" (NSRF 2014–2020) and co-financed by Greece and the European Union (European Regional Development Fund).

Conflicts of Interest: The authors declare no conflict of interest.

References

1. Özgür, Ü.; Alivov, Y.I.; Liu, C.; Teke, A.; Reshchikov, M.A.; Doğan, S.; Avrutin, V.; Cho, S.J.; Morkoç, H. A comprehensive review of ZnO materials and devices. *J. Appl. Phys.* **2005**, *98*, 041301. [CrossRef]
2. Willander, M.; Nur, O.; Zhao, Q.X.; Yang, L.L.; Lorenz, M.; Cao, B.Q.; Pérez, J.Z.; Czekalla, C.; Zimmermann, G.; Grundmann, M.; et al. Zinc oxide nanorod based photonic devices: Recent progress in growth, light emitting diodes and lasers. *Nanotechnology* **2009**, *20*, 332001. [CrossRef] [PubMed]
3. Bhati, V.S.; Hojamberdiev, M.; Kumar, M. Enhanced sensing performance of ZnO nanostructures-based gas sensors: A review. *Energy Rep.* **2020**, *6*, 46–62. [CrossRef]
4. Theerthagiri, J.; Salla, S.; Senthil, R.; Nithyadharseni, P.; Madankumar, A.; Arunachalam, P.; Maiyalagan, T.; Kim, H.-S. A review on ZnO nanostructured materials: Energy, environmental and biological applications. *Nanotechnology* **2019**, *30*, 392001. [CrossRef]
5. Serrà, A.; Gómez, E.; Philippe, L. Bioinspired ZnO-Based Solar Photocatalysts for the Effcient Decontamination of Persistent Organic Pollutants and Hexavalent Chromium in Wastewater. *Catalysts* **2019**, *9*, 974. [CrossRef]
6. Qadir, A.M.; Erdoğan, I.Y. Structural properties and enhanced photoelectrochemical performance of ZnO films decorated with Cu2O nanocubes. *Int. J. Hydrog. Energy* **2019**, *44*, 18694–18702. [CrossRef]
7. Borysiewicz, M.A. ZnO as a Functional Material: A Review. *Crystals* **2019**, *9*, 505. [CrossRef]

8. Kumar, R.; Kumar, G.; Al-Dossary, O.; Umar, A. ZnO nanostructured thin films: Depositions, properties and applications—A review. *Mater. Express* **2015**, *5*, 3–23. [CrossRef]
9. Kumar, M.; Sasikumar, C. Electrodeposition of Nanostructured ZnO Thin Film: A Review. *Am. J. Mater. Sci. Eng.* **2014**, *2*, 18–23. [CrossRef]
10. Kołodziejczak-Radzimska, A.; Jesionowski, T. Zinc Oxide—From Synthesis to Application: A Review. *Materials* **2014**, *7*, 2833–2881. [CrossRef]
11. Singh, G.; Singh, P.S. Synthesis of zinc oxide by sol-gel method and to study its structural properties. *AIP* **2020**, *2220*, 020184.
12. Aldalbahi, A.; Alterary, S.; Almoghim, R.A.A.; Awad, M.A.; Aldosari, N.S.; Alghannam, S.F.; Alabdan, A.N.; Alharbi, S.; Alateeq, B.A.M.; Mohsen, A.; et al. Greener Synthesis of Zinc Oxide Nanoparticles: Characterization and Multifaceted Applications. *Molecules* **2020**, *25*, 4198. [CrossRef] [PubMed]
13. Skompska, M.; Zarebska, K. Electrodeposition of ZnO Nanorod Arrays on Transparent Conducting Substrates—A Review. *Electrochim. Acta* **2014**, *127*, 467–488. [CrossRef]
14. Xu, L.; Guo, Y.; Liao, Q.; Zhang, J.; Xu, D. Morphological Control of ZnO Nanostructures by Electrodeposition. *J. Phys. Chem. B* **2005**, *109*, 13519–13522. [CrossRef]
15. Alev, O.; Sarıca, N.; Ozdemir, O.; Arslan, L.Ç.; Büyükkose, S.; Oztürk, Z.Z. Cu-doped ZnO nanorods based QCM sensor for hazardous gases. *J. Alloys Compd.* **2020**, *826*, 154177. [CrossRef]
16. Lupan, O.; Pauporte, T.; Chow, L.; Viana, B.; Pelle, F.; Ono, L.K.; Cuenya, B.R.; Heinrich, H. Effects of annealing on properties of ZnO thin films prepared by electrochemical deposition in chloride medium. *Appl. Surf. Sci.* **2010**, *256*, 1895–1907. [CrossRef]
17. Boukos, N.; Chandrinou, C.; Giannakopoulos, K.; Pistolis, G.; Travlos, A. Growth of ZnO nanorods by a simple chemical method. *Appl. Phys. A* **2007**, *88*, 35–39. [CrossRef]
18. Haase, M.; Weller, H.; Henglein, A. Photochemistry and radiation chemistry of colloidal semiconductors. 23. Electron storage on zinc oxide particles and size quantization. *J. Phys. Chem.* **1988**, *92*, 482–487. [CrossRef]
19. Seow, Z.L.S.; Wong, A.S.W.; Thavasi, V.; Jose, R.; Ramakrishna, S.; Ho, G.W. Controlled synthesis and application of ZnO nanoparticles, nanorods and nanospheres in dye-sensitized solar cells. *Nanotechnology* **2008**, *20*, 045604. [CrossRef] [PubMed]
20. Zhang, Z.; Yu, H.; Shao, X.; Han, M. Near-Room-Temperature Production of Diameter-Tunable ZnO Nanorod Arrays through Natural Oxidation of Zinc Metal. *Chem. A Eur. J.* **2005**, *11*, 3149–3154. [CrossRef] [PubMed]
21. Sakellis, I.; Giamini, S.; Moschos, I.; Chandrinou, C.; Travlos, A.; Kim, C.-Y.; Lee, J.-H.; Kim, J.-G.; Boukos, N. A novel method for the growth of Cu2O/ZnO heterojunctions. *Energy Proc.* **2014**, *60*, 37–42. [CrossRef]
22. Lupan, O.; Guérina, V.M.; Tiginyanub, I.M.; Ursakib, V.V.; Chowc, L.; Heinrichc, H.; Pauporté, T. Well-aligned arrays of vertically oriented ZnO nanowires electrodeposited on ITO-coated glass and their integration in dye sensitized solar cells. *J. Photochem. Photobiol. A* **2010**, *211*, 65–73. [CrossRef]
23. Pauporte, T.; Bataille, G.; Joulaud, L.; Vermersch, F.J. Well-aligned ZnO nanowire arrays prepared by seed-layer-free electrodeposition and their Cassie–Wenzel transition after hydrophobization. *J. Phys. Chem. C* **2010**, *114*, 194–202. [CrossRef]
24. Zhang, Z.; Yu, H.; Wang, Y.; Han, M.-Y. Aggregation-driven growth of well-oriented ZnO nanorod arrays. *Nanotechnology* **2006**, *17*, 2994–2997. [CrossRef]
25. Li, W.J.; Shi, E.W.; Zhong, W.Z.; Yin, Z.W. Growth mechanism and growth habit of oxide crystals. *J. Cryst. Growth* **1999**, *203*, 186–203. [CrossRef]
26. Ahsanulhaq, Q.; Umar, A.; Hahn, Y.B. Growth of aligned ZnO nanorods and nanopencils on ZnO/Si in aqueous solution: Growth mechanism and structural and optical properties. *Nanotechnology* **2007**, *18*, 115603. [CrossRef]
27. Mbam, S.O.; Nwonu, S.E.; Orelaja, O.A.; Nwigwe, U.S.; Gou, X.-F. Thin-film coating; historical evolution, conventional deposition technologies, stress-state micro/nano-level measurement/models and prospects projection: A critical review. *Mater. Res. Express* **2019**, *6*, 122001. [CrossRef]
28. Kumar, J.; Srivastava, A.K. Band gap narrowing in zinc oxide-based semiconductor thin films. *J. Appl. Phys.* **2014**, *115*, 134904. [CrossRef]
29. Meyer, B.K.; Alves, H.; Hofmann, D.M.; Kriegseis, W.; Forster, D.; Bertram, F.; Christen, J.; Hoffmann, A.; Straßburg, M.; Dworzak, M.; et al. Bound exciton and donor–acceptor pair recombinations in ZnO. *Phys. Status Solidi* **2004**, *241*, 231–260. [CrossRef]
30. Rodnyi, P.A.; Khodyuk, I.V. Optical and luminescence properties of zinc oxide (Review). *Opt. Spectrosc.* **2011**, *111*, 776–785. [CrossRef]
31. Chris, A.J.; Van De Walle, G. Fundamentals of zinc oxide as a semiconductor. *J. Cryst. Growth* **2006**, *72*, 126501.
32. Sakellis, I. Determining the activation volumes in ZnO. *J. Appl. Phys.* **2012**, *112*, 13504. [CrossRef]
33. Kohan, A.F.; Ceder, G.; Morgan, D.; Van De Walle, C.G. First-principles study of native point defects in ZnO. *Phys. Rev. B* **2009**, *61*, 15019. [CrossRef]
34. Guo, B.; Qiu, Z.R.; Wong, K.S. Intensity dependence and transient dynamics of donor–acceptor pair recombination in ZnO thin films grown on (001) silicon. *Appl. Phys. Lett.* **2003**, *82*, 2290. [CrossRef]
35. Liu, W.; Gua, S.L.; Ye, J.D.; Zhu, S.M.; Liu, S.M.; Zhou, X.; Zhang, R.; Shi, Y.; Zheng, Y.D. Blue-yellow ZnO homostructural light-emitting diode realized by metalorganic chemical vapor deposition technique. *Appl. Phys. Lett.* **2006**, *88*, 092101. [CrossRef]
36. Leiter, F.H.; Alves, H.R.; Hofstaetter, A.; Hofmann, D.M.; Meyer, B.K. The Oxygen Vacancy as the Origin of a Green Emission in Undoped ZnO. *Phys. Status Solidi* **2001**, *226*, R4–R5. [CrossRef]
37. Leiter, F.H.; Alves, H.; Pfisterer, D.; Romanov, N.G.; Hofmann, D.M.; Meyer, B.K. Identification of oxygen and zinc vacancy optical signals in ZnO. *Physica B* **2003**, *201*, 340.
38. Liu, M.; Kitai, A.; Mascher, P. Point defects and luminescence centres in zinc oxide and zinc oxide doped with manganese. *J. Lumin.* **1992**, *54*, 35–42. [CrossRef]

39. Reynolds, D.C.; Look, D.C.; Jogai, B.; Hoelscher, J.E.; Sherriff, R.E.; Harris, M.T.; Callahan, M.J. Time-resolved photoluminescence lifetime measurements of the Γ5 and Γ6 free excitons in ZnO. *J. Appl. Phys.* **2000**, *88*, 2152. [CrossRef]
40. Dingle, R. Luminescent Transitions Associated With Divalent Copper Impurities and the Green Emission from Semiconducting Zinc Oxide. *Phys. Rev. Lett.* **1969**, *23*, 579. [CrossRef]
41. Studenikin, S.A.; Golego, N.; Cocivera, M. Fabrication of green and orange photoluminescent, undoped ZnO films using spray pyrolysis. *J. Appl. Phys.* **1998**, *84*, 2287. [CrossRef]
42. Alivov, Y.I.; Chukichev, M.V.; Nikitenko, V.A. Green luminescence band of zinc oxide films copper-doped by thermal diffusion. *Semiconductors* **2004**, *38*, 31–35. [CrossRef]
43. Chandrinou, C.; Boukos, N.; Stogios, C.; Travlos, A. PL study of oxygen defect formation in ZnO nanorods. *Microelectron. J.* **2009**, *40*, 296–298. [CrossRef]
44. Travlos, A.; Boukos, N.; Chandrinou, C.; Kwack, H.-S.; Dang, L.S. Zinc and oxygen vacancies in ZnO nanorods. *J. Appl. Phys.* **2009**, *106*, 104307. [CrossRef]

fibers

MDPI

Review

Mechanical and Dielectric Properties of Aligned Electrospun Fibers

Blesson Isaac [1,*], Robert M. Taylor [2] and Kenneth Reifsnider [2]

[1] Department of Chemical and Radiation Measurement, Idaho National Laboratory, Idaho Falls, ID 83415, USA
[2] Department of Mechanical and Aerospace Engineering, The University of Texas at Arlington, Arlington, TX 76039, USA; taylorrm@uta.edu (R.M.T.); kenneth.reifsnider@uta.edu (K.R.)
* Correspondence: blesson.isaac@inl.gov; Tel.: +1-713-553-4037

Abstract: This review paper examines the current state-of-the-art in fabrication of aligned fibers via electrospinning techniques and the effects of these techniques on the mechanical and dielectric properties of electrospun fibers. Molecular orientation, system configuration to align fibers, and post-drawing treatment, like hot/cold drawing process, contribute to better specific strength and specific stiffness properties of nanofibers. The authors suggest that these improved, aligned nanofibers, when applied in composites, have better mechanical and dielectric properties for many structural and multifunctional applications, including advanced aerospace applications and energy storage devices. For these applications, most fiber alignment electrospinning research has focused on either mechanical property improvement or dielectric property improvement alone, but not both simultaneously. Relative to many other nanofiber formation techniques, the electrospinning technique exhibits superior nanofiber formation when considering cost and manufacturing complexity for many situations. Even though the dielectric property of pure nanofiber mat may not be of general interest, the analysis of the combined effect of mechanical and dielectric properties is relevant to the present analysis of improved and aligned nanofibers. A plethora of nanofibers, in particular, polyacrylonitrile (PAN) electrospun nanofibers, are discussed for their mechanical and dielectric properties. In addition, other types of electrospun nanofibers are explored for their mechanical and dielectric properties. An exploratory study by the author demonstrates the relationship between mechanical and dielectric properties for specimens obtained from a rotating mandrel horizontal setup.

Keywords: electrospinning; aligned nanofibers; molecular orientation; mechanical; dielectric

Citation: Isaac, B.; Taylor, R.M.; Reifsnider, K. Mechanical and Dielectric Properties of Aligned Electrospun Fibers. *Fibers* **2021**, *9*, 4. https://doi.org/10.3390/fib9010004

Received: 17 October 2020
Accepted: 11 December 2020
Published: 6 January 2021

1. Introduction

Many advanced applications can benefit from electrospun materials with superior mechanical and dielectric properties, especially in the fields of composite reinforcement and energy. Aligned electrospun fibers, more specifically, have applications in structural reinforcement of materials and energy storage devices. For these applications, it is paramount to understand the effects of electrospun fiber alignment on mechanical and dielectric properties. Though adding appropriate fillers to the polymers changes mechanical and dielectric properties, better fiber alignment alone improves these properties and keeps the composition uniform throughout. Mechanical and dielectric properties depend on the density and porosity of nanofiber mats, as well as the fiber morphology, including the fiber diameter, and the effect of degree of alignment [1–3]. Therefore, it is necessary to understand the knowledge on both mechanical and dielectric properties of polymer mats together. Mechanical and dielectric properties are among the most important parameters to determine the performance of the polymeric nanomaterials [4,5]. Electrospinning influences both mechanical and dielectric properties of nanofiber membranes [4]. Electrospinning is the process of producing micro- and nanofibers, using a polymer solution with a syringe pump, syringe, needle, collector, and high-voltage power supply. The typical setup of an electrospinning apparatus is either horizontal electrospinning or vertical electrospinning [6]. Figure 1 shows the schematic setup of both types.

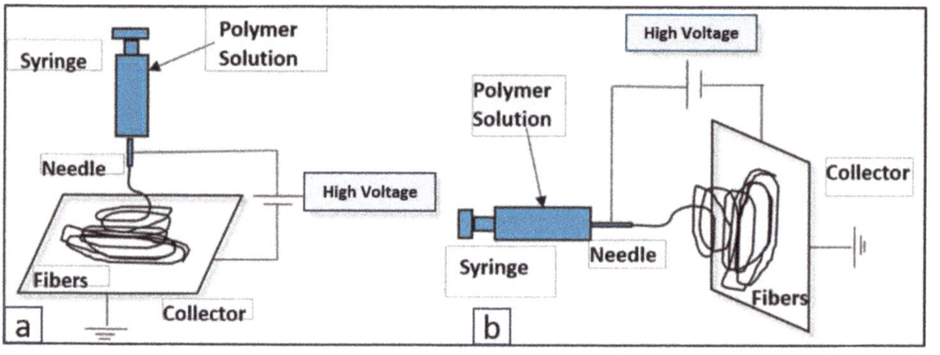

Figure 1. (**a**) Schematics of electrospinning apparatus of vertical setup and (**b**) horizontal setup.

1.1. History of Electrospinning

The idea of electrospinning can be traced back to 1900, when John. F Cooley received the patent for his apparatus for electrically separating the relatively volatile liquid component from the component of relatively fixed substances of composites [7]. Later in 1902, John F. Cooley invented an apparatus for electrically dispersing fluids [8] and William James Morton invented methods of dispersing fluids by the process of separating the volatile components and breaking up the fixed component from composite fluids [9]. Anton Formhals received a patent in the year 1934 for his invention of producing polymer threads, using electrostatic force. In his paper titled "Process and apparatus for preparing artificial threads" [10], solutions of cellulose esters, specifically cellulose acetate were used for spinning. In US Patent No. 2,160,962 (1939), artificial fibers were collected as substantially parallel to each other on a moving collecting device [11]. There he introduced the term "electrical spinning" of fibers. In the spinning process there were difficulties in solidifying the formed fibers. In addition, the as-processed fibers were so sticky that, not only would they stick to the collecting device, but also they would stick to each other. He observed that it was difficult to control the paths of high-speed liquid streams and the corresponding fibers out of it. As shown in Figure 2, fiber direction guide (55 in Figure 2), which consists of shields (57 in Figure 2) to direct the fibers along fixed, predetermined paths toward the collecting electrodes was used. This invention made it possible to obtain smooth, continuous, compact, and coherent fiber bands composed of heterogeneous filaments arranged substantially parallel to each other.

Zhang et al. (2016) reported that different nanofiber production methods include vapor growth, arc discharge, laser ablation, and chemical vapor deposition [12]. These processes are very expensive because of low product yield and high equipment cost. However, electrospinning employs a top-down engineering approach, which can produce fibers with diameters ranging from 10 nm to 10 μm, from a polymer solution, under the application of an electrostatic force [13,14]. These fibers have a high surface area to volume ratio, high porosity, and tunable porosity [6]. According to Luo et al. (2012), there are various spinning techniques available for producing micro and nanofibers [14]. Solution electrospinning compared to melt electrospinning requires a solvent. The melt electrospinning method uses a molten polymer, but the absence of solvent excludes the effect of solvent properties on the fiber formation. In emulsion electrospinning, two immiscible fluids are used as in food-processing [15]. Magnetic electrospinning and near-field electrospinning are good examples of interdisciplinary technological convergence between magnetism and electric potential methods. Dip-pen nanolithography with traditional electrospinning can also be used, but the alignment of fibers is not satisfactory [14].

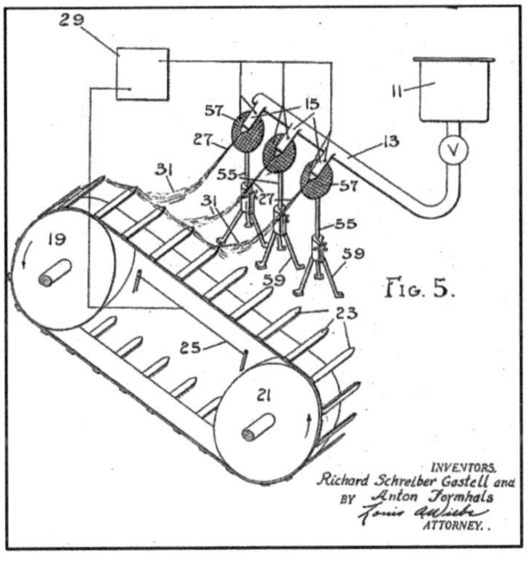

Figure 2. Electrospinning setup by A. Formhals [11].

1.2. Working Principle of Electrospinning

The working principle for electrospinning is shown in the Figure 3. A sufficiently high voltage is applied at the location of the liquid droplets formed at the tip of the needle. The local body of the liquid becomes charged. Electrostatic repulsion counteracts the surface tension. Thus, the droplet is stretched, and at a critical point, a stream of liquid erupts from the surface. The point of eruption is called a Taylor Cone. Sir Geoffrey Taylor developed the equation which shows the relationship between the critical voltage and the surface tension as shown in Equation (1) [16,17].

$$V_c^2 = \frac{4H^2}{L^2}\left(ln\frac{2L}{R} - \frac{3}{2}\right)(0.117\prod \gamma R) \quad (1)$$

where V_c is the critical voltage, H is distance between the needle tip and the collector, L is the length of the needle with radius R, and γ is the surface tension of the liquid (units: V_c in kilovolts; H, L, and R in cm; and γ in dyne per cm). Afshari (2017) showed that electrostatic forces play a key role on the electrospinning of polymer solutions [18]. As such, the Coulomb's force is considered as the driving factor for better design. In Equation (2) shown below, F is the Coulomb's force, k is the constant of proportionality, q_1 and q_2 are charges, and r is the distance between the charges. In principle, the smaller the distance, the greater the electrostatic force on a charged particle.

$$F = k\frac{q_1\, q_2}{r^2} \quad (2)$$

When an electric field is applied, the liquid jet ejected from the tip of the nozzle/needle travels on a straight line for a short distance. The diameter of the jet, in the straight line, decreases monotonically with the distance from tip, after that a radially outward bending instability happens. The electrostatic force from the charge carries with the jet causes the jet to continue to elongate as it coils and the thin fluid jet solidifies into nanofiber [19].

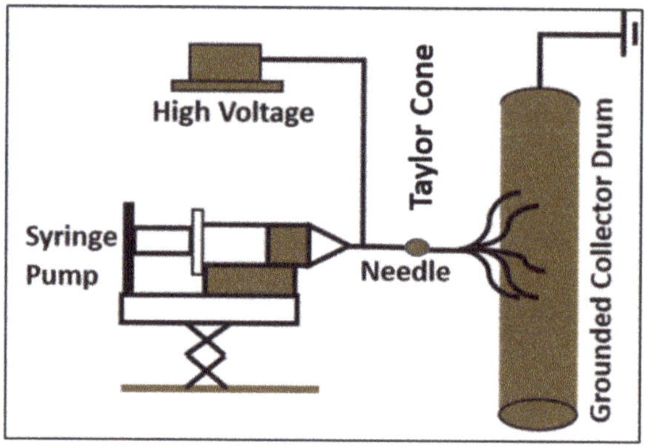

Figure 3. Working principle of electrospinning.

1.3. Applications of Electrospun Fibers

The typical applications of electrospun fibers include filtration, energy, structures, biomedical, textiles, and others [6] as shown in Figure 4. Other applications include optical and chemical sensors, textiles, reinforcement of composites, health care, and defense and security. Electrospun fibers are projected to play an important role in the development of air filtration, energy storage devices, super-capacitors, and rechargeable batteries [20–26].

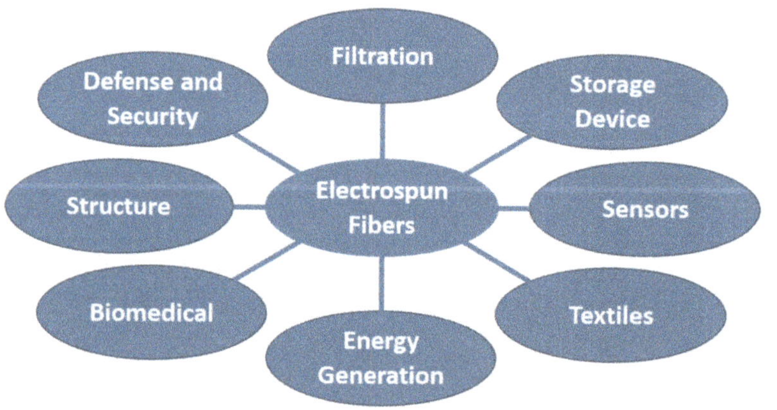

Figure 4. Applications of electrospun nanofibers.

The applications of nanofiber mats in the reinforcement of nanocomposites are discussed by Huang et al. (2003), who executed mechanical characterization of nanofibrous membranes of various polymers and examined their potential applications [17]. Nanofibers can have better mechanical properties than microfibers and therefore superior structural properties can be anticipated. Jiang et al. (2018) provided an overview of nanofiber composite application [27]. Bergshoef and Vancso (1999) showed that smooth nylon-4, six electrospun fibers with diameters in the range of 30–200 nm can be produced from formic acid solutions. These fibers demonstrated reinforcement of transparent composites with an epoxy matrix [28]. Highly porous nanofibers with pore interconnectivity and relatively uniform pore distributions improve membrane performance in the application of desalination (water filtration) [29]. The large surface area of the constituent fibers provides high functionalization capability and mechanical bonding to limit delamination between laminae [30]. Biomedical applications include tissue engineering scaffolds, wound dressing, drug delivery [31], and creation of artificial blood vessels. The non-woven nanofibrous mats produced by electrospinning techniques mimic the extracellular matrix components.

Some important issues and challenges in the 21st century are addressed by using electrospun fibers in the domains of tissue regeneration, energy conversion and storage, and water treatment. Large surface areas, high porosity, and the unique mat structure of electrospun nanofibers have provided improvements over the last decade in the fields of tissue regeneration (skin [32–34], nerve [35–37], heart [38–40], and bone [41–43]), energy conversion and storage (solar cells [44–47], fuel cells [48–51], and batteries [52–55]), and water treatment (adsorption [56–58], photocatalysis [59–61], and filtration [62–64]). Potential applications and promising advantages are overviewed by Bhardwaj and Kundu (2010), who highlighted more than 200 polymers that are electrospun for various applications. Teo and Ramakrishna (2006) gave a detailed review on electrospinning design and nanofiber assemblies [65].

1.4. Recent Review Papers on Electrospun Nanofiber

Table 1 lists twelve recent review papers on electrospun fiber applications and characterizations. Of the papers considered, reviews of applications dominate the literature, are a few on mechanical, energy, medical, and processing characterizations. However, there is little work found in the literature on dielectric and mechanical properties together that should contribute to both composite reinforcement and energy applications.

Table 1. Recent review papers on electrospun nanofibers.

Authors	Year	Main Criteria of Review Papers
Huang et al.	2003	Processing, structure, characterization, applications, modeling and simulation, and different polymers in solution and melt form [17]
Pham et al.	2006	Tissue engineering (scaffolds) [66]
Bhardwaj and Kundu	2010	Polymers, parameters, melt electrospinning, and applications [6]
Luo et al.	2012	Scale-up challenges and applications [14]
Shuakat et al.	2014	Nanofiber yarns and nanofiber alignment [67]
Shi et al.	2015	1D nanomaterials have high surface-area-to-volume (specific surface area), high aspect ratio, and high pore volume. Well-aligned and highly ordered are suitable for energy harvesting and storage devices. More advantageous than conventional materials [68]
Ahmed et al.	2015	Desalination [29]
Zhang et al.	2016	Energy storage [12]
Peng et al.	2016	Tissue regeneration, energy conversion and storage, and water treatment [23]
Shekh et al.	2017	Water purification [69]
Zhang et al.	2018	Food packaging [15]
Li et al.	2019	Electrical and mechanical performance of polymer nanocomposites [70]

1.5. Parameters and Parameter Optimizations

Important parameters that affect the quality of electrospun fibers formed from polymer solutions can be categorized as solution-specific parameters, process-specific parameters, and environmental-specific parameters [6,17,71].

(a) Solution parameters: The solution-specific parameters include viscosity, polymer concentration, surface tension, conductivity, and evaporation rate of solvent [18,72–76]. It is observed that low viscosity is typically responsible for bead generation and significant increase in fiber diameter. A similar conclusion was made on polyacrylonitrile/dimethylformamide (PAN/DMF) solution where beads were easier to form at low concentration of 5 wt.% than that formed at higher concentration of 7 wt.% [77,78]. Typically, viscosity and concentration are directly proportional to each other [79].

Additionally, polymer concentration directly controls fiber diameter [79]. In general, an increase in fiber diameter can be achieved by increasing the polymer concentration. Higher surface tension causes bead formation and reduced surface tension favors smooth fiber formation [80].

(b) Process parameters: Applied voltage, distance between the nozzle tip and collector, rotating speed of the collector (if drum is used), and solution feed rate are the parameters that are regarded as process specific [18,81–83]. In general, fiber diameter can be reduced by increasing applied voltage and vice versa. If the applied voltage reaches a critical value, a charged jet initiates the electrospinning process. This critical voltage is closely related to surface tension of the solution. Lee et al. (2003) reported that there was a linear relationship between voltage applied and surface tension of polystyrene (PS) dissolved in a mixture of tetrahydofuran and DMF [84]. The distance between the tip and the collector mainly controls fiber solidification because a minimum distance is required to allow the fibers sufficient time to dry before reaching the collector. Distances that are too close or too far can cause beads to form. Fang et al. (2010) studied 7 wt.% PAN/DMF electrospun at 2–10 cm away from nozzle tip. The experiments concluded that beads were producing until the distance reached 7 cm [78]. Longer distance between nozzle tip and collector produced bead free fibers.

(c) Environmental parameters: Humidity and temperature are treated as environment-specific parameters [18,85,86]. According to De Vrieze et al. (2009), the evaporation rate increases with increase in temperature [87]. Moreover, the viscosity of solution generally decreases with an increase in temperature. As the humidity increases, the average fiber diameter increases. Parameter optimization: Formation of nanofibers involve many input parameters, as mentioned above, to evaluate outputs such as fiber diameter, tensile strength, modulus, and dielectric properties of nanofibers. Parameter optimization helps to achieve desired outputs by tailoring the input parameters. One among the many mathematical modeling techniques for parameter optimization is Design of Experiment (DoE), which is an approach that helps to find the relationship between different inputs over outputs. Parameter optimization based on applied voltage and concentration has been studied by using the DoE approach by Gu et al. (2005) [88]. The study concluded that concentration of solution played an important role to the diameter of nanofibers. Gu et al. (2005) used two factors and four and three respective levels for finding average fiber diameter. Senthil and Anandhan (2005) examined three variables and seven, four, and three respective factor levels for finding the average fiber diameter [89]. Isaac et al. (2018) used DoE approach with two factors and three levels for optimizing the two outputs, namely, specific dielectric constant and specific mechanical strength [90,91]. A mathematical modeling, including the leaky dielectric model which describes the deformation of a Newtonian drop in an electric field and whipping model which depicts the interaction between the electric field and fluid properties for electrospinning processes, has been portrayed by Rafiei et al. (2013) [92]. Ismail et al. (2016) developed a model for stable region and unstable region in the jet propulsion stream for predicting the fiber diameter [93]. Rafiei et al. (2014) modeled and simulated viscoelastic elements for jet propulsion to predict and improve control of nanofiber diameter [94]. Modeling electrospinning of nanofibers for short-range and long-range electrostatic interactions, using a discrete slender model, was conducted by Kowalewski et al. (2009) [95]. The whipping instability in the unstable region of the electrospinning jet propagation has been studied in three polymeric solutions by Kowalewski et al. (2005) [96]. The fiber gets stretched into fractions of initial diameter at the instability region. Ghaly (2014) modeled the electrospinning jet with an inkjet printer technique, using computer-aided fluid/multi-physics/multi-phase flow simulations in COMSOL multiphysics software [97].

2. Molecular Orientation and System Configurations of Nanofibers

Two key factors affecting mechanical and dielectric properties are (a) molecular orientation due to elongation of fibers on the periphery of the rotating mandrel [98] and (b) system configuration improvement for obtaining improved properties due to better alignment [99]. Other properties, such as thermal and electrical properties [100] are also often improved by alignment.

2.1. Molecular Orientation of Nanofibers

High orientation of polymer molecular chains along the fiber axis and aligned electrospun fibers have important consequences in the field of carbon fiber-reinforced nanocomposites. The electrospun fibers are generally stronger than traditional fibers because of their higher orientation of macromolecular polymer chains along the fiber axis. The polymer jet under the influence of an electrostatic field experiences a high degree of molecular orientation due to high elongation strains and shear forces. As explained later, in Section 2.2, System Configuration to Align Fibers, optimal speed of fiber collecting drum brings about better alignment. In addition, optimal speed of collecting drum causes maximum molecular orientation. Beyond the optimal speed, the orientation can decrease slightly. According to Fennessey and Farris (2004), twisted yarns of higher degree of molecular orientation resulted in better mechanical properties [99]. The degree of orientation can be quantified by the X-ray diffraction analysis of the samples. The nitrile group in PAN is oriented in approximately perpendicular to the draw direction. The absorbance of perpendicular polarization showed nitrile-stretching vibration with strong dichroism, and therefore better orientation. A twist angle of 11° in as-spun PAN fiber improved the initial modulus and ultimate strength of 2.6 GPa and 56 MPa, respectively, to 2.2 times and 2.9 times, respectively. Molecular orientation results in better mechanical properties in general, Young's modulus in particular, of the resulting carbon fibers [101,102]. Baji et al. (2010) studied the effects of electrospun polymers on oriented morphology and tensile properties. The lower the diameter of the fibers, the higher the modulus and strength of the fibers. They observed that finer fibers have enhanced properties because of gradual ordering of molecular chains and increase in crystallinity [103]. Baji et al. (2010) noticed that the modulus and tensile properties of polycaprolactone (PCL) fibers increased significantly when the fiber diameter was reduced to below 500 nm. The molecular orientation improves gradually as the fiber diameter is reduced. Moreover, Beese et al. (2013) concluded that electrospun PAN fibers have better mechanical properties at lower diameters [104]. Arshad et al. (2011) observed that the strength of the carbonized nanofibers at 800 °C increased by 100% when the diameter was reduced from 800 to 200 nm [105]. For composite applications, a decrease in diameter of fiber at the nanoscale level can improve mechanical properties as the specific reinforcement area per unit mass increases. Uyar et al. (2009) observed self-aligned bundled fibers of polyphenylene-g-polystyrene/poly (a-caprolactone) (PP-g-PS/PCL) when blended with polystyrene (PS) or polymethyl methacrylate (PMMA). This is because of the unique molecular architecture of PP-g-PS/PCL and its interaction with PS or PMMA [106].

2.2. System Configuration to Align Fibers

In addition to molecular orientation, physical alignment of electrospun fibers contributes to the production of high strength/high toughness fiber reinforced composites [12]. Among the many ways to produce aligned fibers by using an electrospinning technique, drum collection and rotating disk collectors are the two most popular designs, as shown in Figure 5a,b [65,107–111].

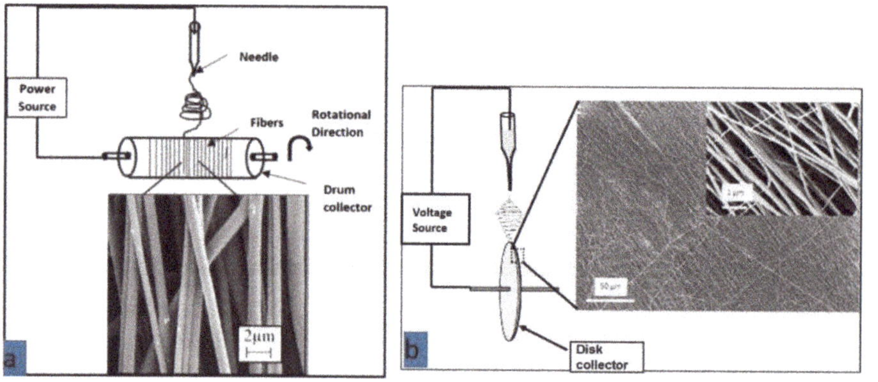

Figure 5. (**a**) Rotating drum collector. (**b**) Rotating disk collector [103].

In the drum collector method, the drum collector rotates at high speed and the deposited fiber diameter can be controlled based on the rotational speed of the drum [112–114]. The linear speed at the surface of the rotational drum should match the evaporation rate of the solvent so that fibers are deposited and taken up on the surface of the drum. At a rotational speed less than the fiber take-up speed, randomly oriented fibers are obtained on the drum. At higher speed, the fiber take-up velocity breaks the fiber, and continuous fibers are not collected. Therefore, the optimal speed of the rotating drum that matches the evaporation rate is required for maximum alignment [115].

In the rotating disk collector, higher alignment is possible but the production rate is lower because fibers are effectively deposited at only a small area at the disk edge. Theron et al. (2001) reported a conical and an inverted conical instability region of polyethylene-based polymer nanofibers. These finer fibers with diameters ranging from 100 to 300 nm got aligned and wound on a sharp edge disk wheel-like bobbin [116].

Other collection methods that are suitable for fiber alignment are the parallel conductor method, the wire drum collector method, and the wheel rotor collector method as shown in Figure 6a–d. In the parallel conductor method [117–120], the length of aligned fibers is restricted by the distance between conductive stripes as shown in Figure 6a. Jalali et al. (2006) reported the fundamental parameters affecting the uniaxially aligned PAN nanofibers. The best alignment of nanofibers with a specific gap distance depends on concentration, voltage, and tip to collector distance. As shown in Figure 6b, a bundle collector is moved across the gap to another side for depositing bundle of nanofibers. The best alignment was formed between 10 and 15 wt.% solutions. Uniaxially aligned fibers formed had an aspect ratio (l/d) of higher than 5000 and these fibers are useful in composite reinforcement application. Fryer et al. (2018) studied the effect of alignment on fiber modulus, using the electrostatic gap method. Aligned polyethylene oxide (PEO) fibers have a higher modulus than the non-aligned fibers of similar diameter [121]. Cai et al. (2017) provided an insight in to fabricating ultra-long polyvinylidene fluoride (PVDF) fibers. Here, parallel conducting U-shaped collectors are used to fabricate fibers [122]. According to Lei et al. (2018), more than a meter long aligned PVDF nanofibers were fabricated by using gap electrospinning, where the needle is connected to positive power supply and the parallel plates are connected to the negative power supply [123]. Yang et al. (2007) demonstrated a method that generates parallel fibers, using magnetic-particle-doped polymers in two parallel placed magnets. The magnetic field guides the magnetized electrospun polyvinyl alcohol (PVA) fibers to align in a parallel fashion [124]. Park and Yang (2011) built uniaxial aligned PCL fibers by introducing an inclined gap into dual collectors that consisted of two conductive stripes which were arranged vertically and horizontally [125]. Dabirian et al. (2009) used a hollow metallic cylinder with needle placed at the center of the cylinder. Fibers produced by this method are claimed to be well aligned and spread over large area [126]. Next, in the wire drum collector method, shown in Figure 6c, fibers are deposited without the need for high speed rotation [126]. However, aligned thick films are not possible with this method. Finally, a wheel rotor collector, as shown in Figure 6d, provides elongation strain and therefore more strength to the fibers. However, the many electrodes on the rotating wheel complicate apparatus design [127]. The limitations of other methods imply that drum collection is more likely to be scalable to commercial capacities than other electrospinning methods for fiber alignment. Despite the simplicity of the electrospinning methodologies, industrial applications are relatively rare due to low fiber throughput for existing fiber collection methods. This throughput limitation could be addressed with larger drum sizes and other innovations.

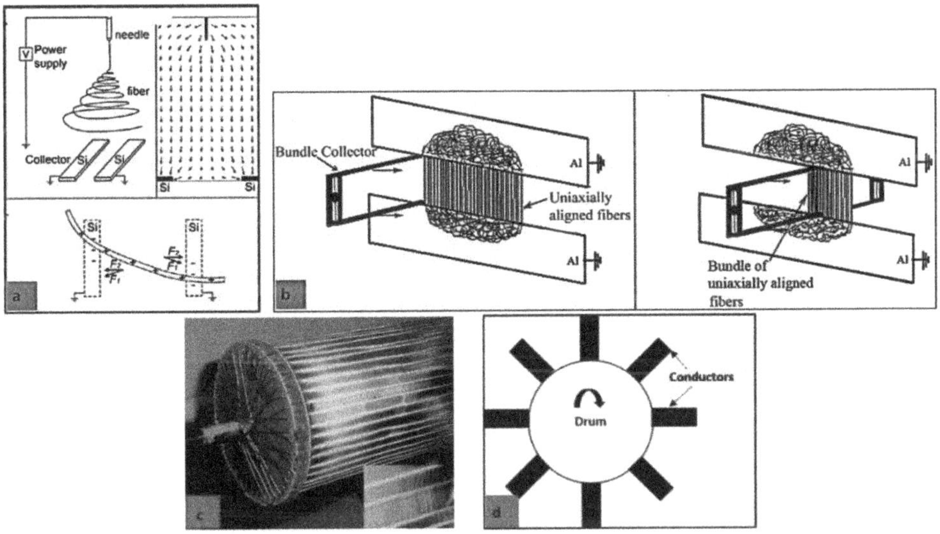

Figure 6. (**a**) Parallel conductor stripes method (reprinted with permission from Reference [128]. Copyright (2003) American Chemical Society). (**b**) Uniaxially aligned nanofibers [129]. (**c**) Wire drum collector (reprinted with permission from Reference [130]. Copyright (2004) American Chemical Society). (**d**) Wheel rotor collector.

In addition to the design methods mentioned above, there a few unconventional design configurations for aligning fibers. Grasl et al. (2013) developed a technique, using two parallel rotatable auxiliary electrodes applied with time-varying square wave potential, which led to aligned fiber-deposition of PEO [131]. Lei et al. (2017) used a collecting system consisting of insulating hollow cylinder and grating-like electrodes for aligning PVDF fibers, using whipping instability [131]. Khamforoush and Mahjob (2011) used a modified rotating jet method for aligning fibers. The degree of alignment enhanced by more than two times and the average amount of produced fiber is 40% more than that of the simple rotating jet method [132,133].

3. Mechanical and Dielectric Properties of Nanofibers

Carbonaceous materials such as carbon black, fullerene, carbon nanotubes, carbon nanofibers, and graphene extend the functionalities of polymers from lightweight and cost-effective to new applications such as electrically and thermally conductive, electromagnetic shielded, etc. With ever-increasing utilities of multifunctional polymer composites applicable in the electronics, sensors, energy, automobile, and aerospace industries, the mechanical properties and electrical are among the two most important parameters to determine the performance of polymeric nanocomposites [70].

3.1. Mechanical Properties

Among many nanofibers that have mechanical properties, PAN nanofibers are widely preferred because of their excellent tensile strength and modulus. In this section, mechanical properties of PAN nanofibers (in general) and carbon nanofillers are discussed. The mechanical properties of electrospun nanofibers, including PCL, polyvinyl pyrrolidone (PVP), and PEO, are also reviewed. Post-treatment techniques, such as drawing and annealing processes, are discussed, since these post-treatments can improve molecular orientation and crystallinity [134]. Moreover, PAN nanofibers in composite applications and other types of nanofibers are discussed in the end.

3.1.1. PAN Nanofibers and Carbon Fillers

Edie (1998) reported that PAN-based carbon fibers have higher tensile strengths and reasonable tensile moduli compared to pitch-based carbon (micro-size) fibers [135]. PAN precursor is used for carbon nanofiber (CNF) production in nanocomposite structures [136] and in energy storage devices [137] due to their good electrospinning property, high carbonization yield, excellent nanostructure, ultrahigh specific surface area, good electrical conductivity, and stability. PAN is soluble in polar solvents like dimethylformamide (DMF), dimethylsulfone ($DMSO_2$), dimethylsulfoxide (DMSO), and dimethylacetamide (DMAc) [13]. Among organic solvents, DMF and DMSO are known to be good solvents of PAN and for production of high-performance PAN fibers, DMSO is preferred [138].

Chawla et al. (2017) observed that carbon nanofibers obtained from hot-drawn samples demonstrated strength as high as 5.4 GPa and modulus 287 GPa as shown in Figure 7a,b. Here, 400 nm PAN nanofibers reached a maximum strength (5.4 GPa) after hot-drawn and carbonized at 1100 °C [139].

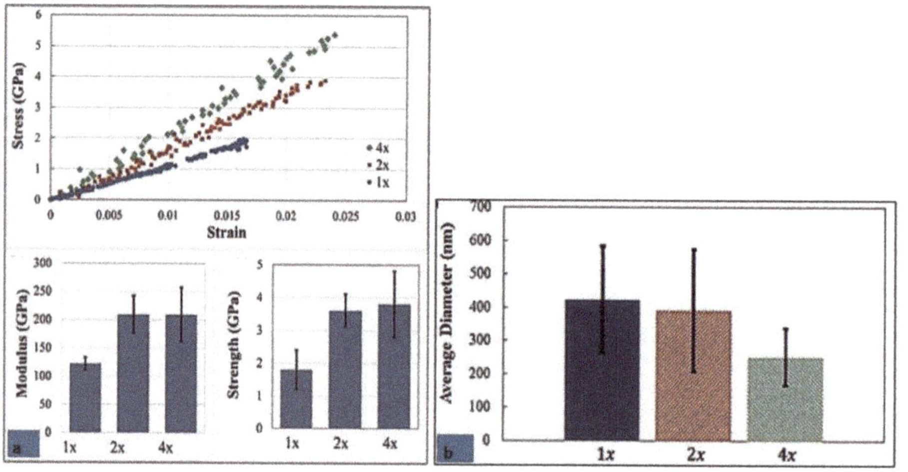

Figure 7. (**a**) Typical stress–strain curve and (**b**) the average fiber diameter of carbon nanofibers (CNFs) [139].

As shown in Figure 8a, the thinnest fiber had the highest strength when compared to other thicker fibers. Moreover, it is evident from Figure 8b that the thinnest fiber had the highest modulus. Papkov et al. (2013) [140] studied simultaneous improvement in strength, modulus, and toughness in ultrafine as-spun PAN electrospun nanofibers as shown in Figure 8a,b. A reduction of as-spun PAN nanofiber diameter from 2.8 µm to 100 nm resulted in higher modulus, strength, and toughness. The 100 nm annealed PAN fibers showed a modulus of 48 GPa and strength of 1.75 GPa. This study recorded dramatic increase in strength and modulus for nanofibers finer than 200–250 nm.

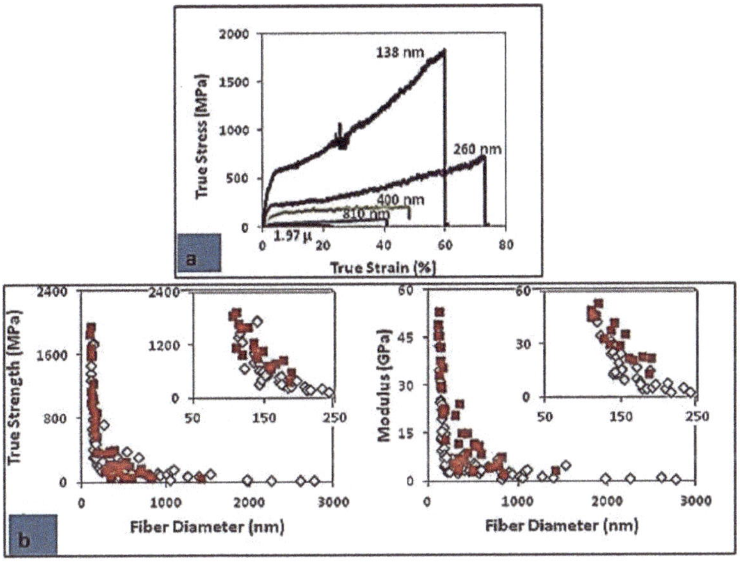

Figure 8. (**a**) Stress–strain graph of as-spun polyacrylonitrile (PAN) nanofibers. (**b**) Strength and modulus of annealed nanofibers. (Reprinted with permission from Reference [140]. Copyright (2004) American Chemical Society).

Beese et al. (2013) [104] compared the result with Arshad et al. (2011) [105] and showed the dependence of fiber diameter on strength and modulus as shown in Figure 9. As the fiber diameter decreases, the tensile strength and modulus increase. Beese et al. (2013) observed that individual PAN nanofibers with 108 nm in diameter, heat treated at 800 °C, showed a maximum modulus of 262 GPa and strength of 7.3 GPa.

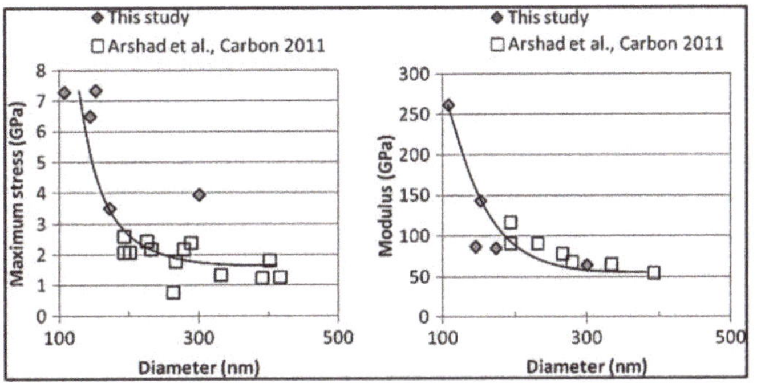

Figure 9. Strength and modulus of individual carbon nanofiber obtained from PAN nanofiber mat [104].

Arshad et al. (2011) [105] showed that individual PAN nanofibers with 9 wt.% and other given parameters had the stress–strain relationship shown in Figure 10. The plot shows a linear relationship until 125 MPa and thereafter a strain hardening region where crystallinity occurs. The maximum ultimate strength was found with 1 kV/cm and 430 nm in diameter. Arshad et al. demonstrated that increase in carbonization temperature in CNFs monotonically increases elastic modulus, while highest strength of CNFs was observed at a carbonization temperature of 1400 °C. This study revealed the fiber diameter and carbonization temperature had effect on strength and modulus of PAN nanofibers.

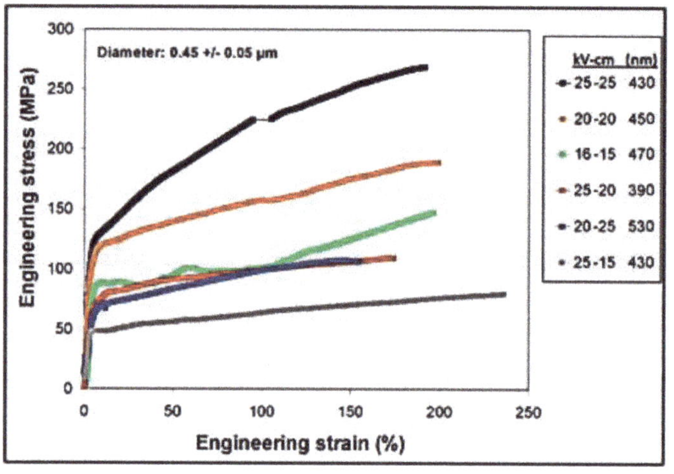

Figure 10. Stress–strain relationship of individual PAN nanofibers [105].

Wan et al. (2015) [1] reported that tensile strength of carbon nanofiber mats can be approximately estimated as the sum of individual nanofibers as shown in Figure 11. The nanofiber mat reached their maximum strength where the curves dropped sharply, which means the majority of nanofibers break simultaneously. Wan et al. (2015) showed that a quantitative relationship exists between the tensile strength of a nanofiber mat and that of individual nanofibers as in Figure 12. The tensile strength of nanofiber mat is described by Equation (3).

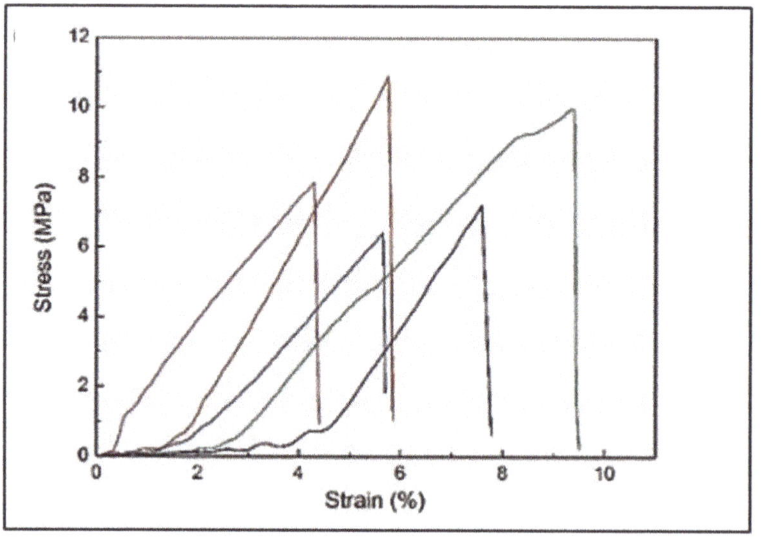

Figure 11. Stress–strain curve of carbon nanofiber mats [1].

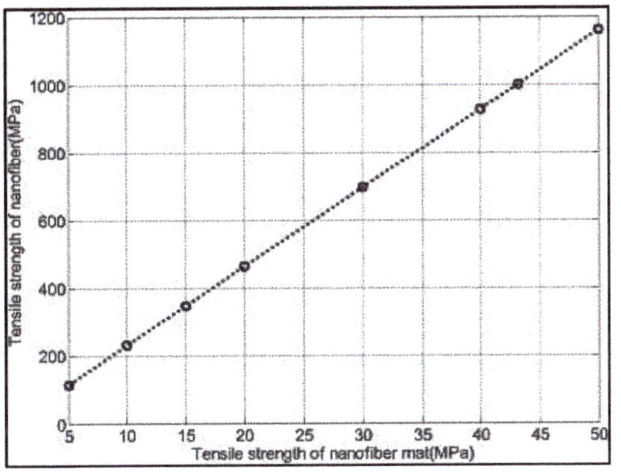

Figure 12. Predicted tensile strength of individual nanofiber vs tensile strength of nanofiber mat [1].

$$\sigma = \frac{(1-P)(\theta + sin\theta cos\theta)^2}{2\pi sin\theta}\sigma_f \tag{3}$$

where σ_f is the fiber strength and θ is the diagonal angle of fibers to the longitudinal axis of loading. The value of θ can be obtained from L = D cosθ and W = D sinθ, where L, W, and D are the length, width, and diagonal of rectangular tensile testing specimen. According to Wan et al. (2015), tensile strength is also a function of the porosity of a nanofiber mat as shown in Figure 13. Specific tensile strength is calculated by using Equation (5). Fiber volume is determined by using Equation (4).

$$\sigma_{sp} = \frac{\sigma}{\rho(1-P)} \tag{4}$$

ρ is the fiber density and P is the porosity.

$$V_f = V_m(1-P) \tag{5}$$

V_m is the volume of mat and P is the porosity.

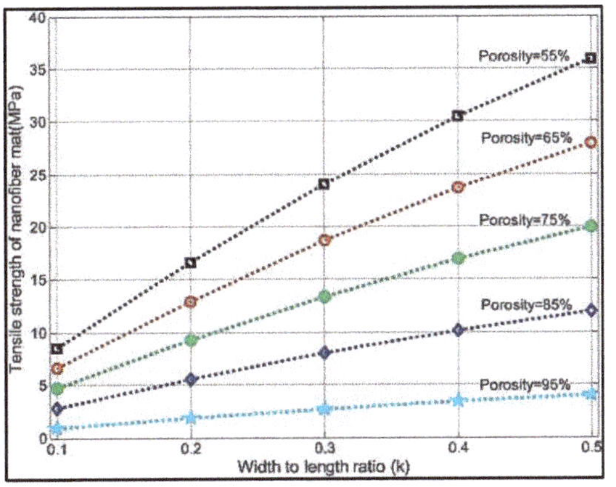

Figure 13. Tensile strength as a function of porosity [1].

Zhang et al. (2016) reported that carbon nanofibers are used in reinforcement of nanocomposites [12]. A carbon nanotube (CNT) is one of the allotropes of carbon. According to Naebe et al. (2010), electrospinning of CNT/polymer has been shown to induce alignment of nanotubes within the matrix. Electropsun CNT/polymer nanofibers showed significant improvement in fiber strength, modulus, and conductivity [141]. One dimensional CNF and CNT have high aspect ratios (typically over a few hundred) that enable them to form a conductive network and they possess excellent mechanical properties. According to Zhang et al. (2016), CNF, compared to CNT, exhibits good dispersion and low fabrication cost. CNT has excellent electrical conductivities, a large surface area to volume ratio, and structural stability [12]. As noticed by Chung (2016), CNT and CNF are difficult to disperse and bond relatively weakly in a matrix. The large area of interface associated with small diameter CNT and CNF aggravate the issue [142]. Addition of small amount of CNT's in PAN precursor nanofibers can improve graphitic order and mechanical properties of carbon nanofibers [143,144]. Both single-walled carbon nanotube (SWCNT) and multi-walled carbon nanotube (MWCNT) [145] can be used as fillers in PAN to enhance molecular orientation. Two types of MWCNT are possible; one is long MWCNT and the other is short MWCNT. Short MWCNTs are 0.5–2 μm in length and the diameter could be 30–50 nm. Graphene is another nanofiller used as a reinforcing material. Compared to CNT and graphene, CNF is considered as a promising reinforcement material due to its relatively high mechanical performance and low fabrication cost. Adding fillers to precursor nanofibers improves tensile strength and tensile modulus as evident in the session, drawing process below.

Drawing Process

Hot drawing of polymers is achieved by drawing the precursors at temperature above the glass transition temperature. Chawla et al. (2017) successfully carried out hot-drawing of PAN electrospun ribbons to enhance molecular orientation. PAN nanofiber ribbons were hot-drawn to two times and four times their original length. These ribbons were further stabilized at 250–300 °C and then carbonized at 1100 °C for 1 h. The tensile modulus and strength of four times hot-drawn ribbons showed maximum values of 287 and 5.4 GPa, which are 71% and 111% increment respectively as compared to as-spun electrospun CNFs. Cai and Naraghi (2019) studied the templating effect of functionalized SWCNTs in CNFs and the contribution of that to mechanical properties of CNFs. To enhance the packing of polymer chains of the precursor around CNT's, PAN as-spun electrospun fibers were subjected to thermomechanical processing (hot-drawing). The MEMS-based single-nanofiber mechanical testing result showed a strong relationship between the modulus, strength improvement and hot-drawing process. The average tensile strength and modulus of CNF/SWCNT were measured to be 7.6 ± 1.72 and 268 ± 29 GPa respectively [146]. Polyamide (PA) fibers are undergone post-drawing process to obtain moderate molecular orientation and crystallinity. After the post-drawing process, PA is found to be with good mechanical strength and abrasion resistance [134]. Yu et al. (2020) revealed enhanced effect of graphene oxide (GO) in PAN nanofiber yarns. The alignment of PAN chains and GO in nanofibers was enhanced by hot-drawing which resulted in increased orientation induced crystallization [147]. Peng et al. (2019) experimentally found that hot-drawn polyethylene fibers with decreasing fiber diameter, Young's modulus increase rapidly. That is due to the fact that chain orientation parameter increases with increasing hot-drawing ratio [148]. Inai et al. (2006) reported that higher rotational speed of fiber collecting disc and post-processing such as annealing and hot-drawing on molecular structure of PLLA nanofibers improved crystalline structure orientation and mechanical strength of fibers. Better mechanical properties were found at higher hot-drawing ratio [149]. Isotactic polypropylene (iPP) nanofibers with the diameter range of 75–375 nm were made from the blends of cellulose acetate butyrate (CAB) and iPP with a ratio of 97.5–2.5. The hot-drawn nanofibers with the ratio of 25 resulted in lower crystallinity than that of bulk iPP. The increase in the amount of CAB in blends gave rise to higher crystallinity in iPP fibers [150].

Cold drawing of polymers is normally done below the glass transition temperature. Cold-drawn nanofibrillated cellulose nanopaper increases its modulus and strength from

10 GPa and 185 MPa to 24.6 GPa and 428 Mpa, at the draw ratio of 1.6 [151]. Cold-drawn blend fibers of PVA and PTFE increased the degree of crystallinity in PTFE/PVA fibers [152]. The addition of hydroxyethyl cellulose (HEC) in cellulose nanofiber improved mechanical property. An aqueous solution of low concentration cellulose nanofiber with HEC promoted nanofiber alignment which was further improved by cold drawing [153].

3.1.2. Other Electrospun Nanofibers

In this session, Alarifi et al. (2009) reported the mechanical property of a PAN-derived carbon nanofiber composite and the effect of molecular orientation along the fiber direction [154]. The study placed electrospun carbonized PAN nanofibers on a stacking sequence of 0, 45, −45, and 45 to create a laminate of ten plies. The tensile testing of PAN-derived carbon nanofiber composites revealed that they possess a high elastic modulus due to stabilization at a high temperature (280 °C), for 1 h, in an oxygen atmosphere. Thermogravimetric analysis (TGA), dynamic mechanical analysis (DMA), thermomechanical analysis (TMA), and differential scanning calorimetry (DSC) analyses confirmed that the nanofibers were crystalline and had good mechanical and thermal properties. Thinner nanofibers have larger surface-area-to-volume ratios. Therefore, thinner fibers have better mechanical integrity between the matrix and the surface of the reinforcing agent for effective load transfer in composites. Baji et al. (2010) proved that as the PCL fiber diameter reduces, the tensile strength and modulus increase [103]. Figure 14a shows the tensile strength and modulus versus fiber diameter. For fiber diameters greater than 2 μm, both tensile modulus and tensile strength appear not to change with diameter. The degree of crystallinity increases gradually as the PCL fiber diameter is reduced as shown in Figure 14b. As for polymers, in general, an increase in the degree of crystallinity increases the density, stiffness, strength, and toughness [155]. Huang et al. (2016) showed that with the addition of conductive filler materials, the diameter of fibers reduced due to increasing composition of PVP/cellulose nanocrystal (CNC)/silver particle [156]. The addition of CNC increased the tensile strength.

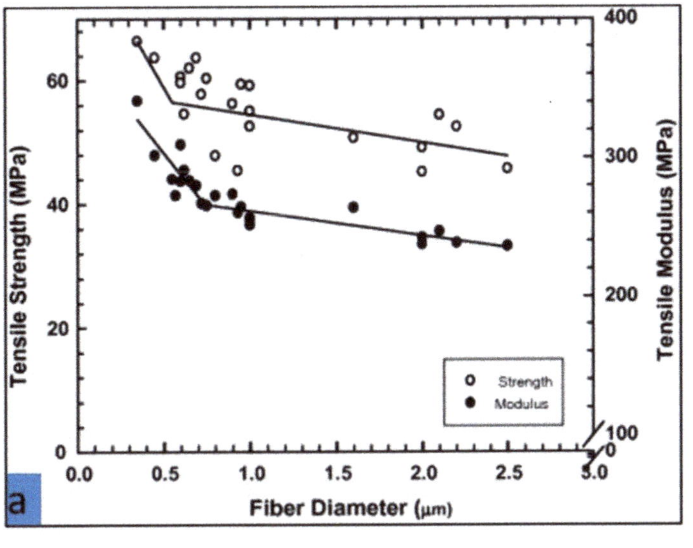

Figure 14. *Cont.*

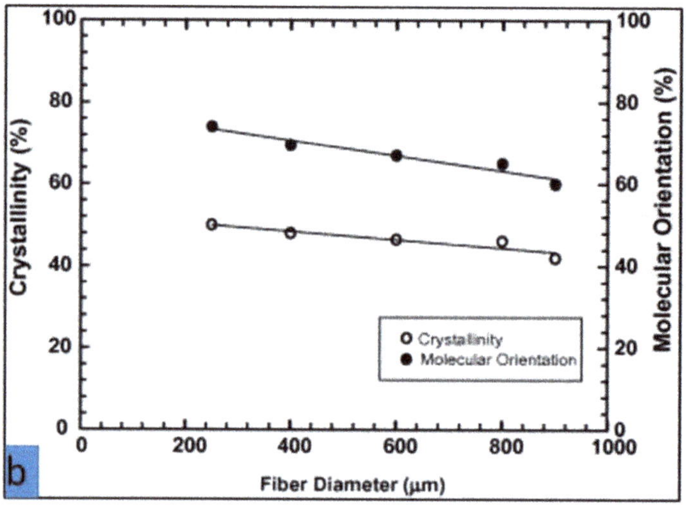

Figure 14. (**a**) Tensile strength/modulus vs fiber diameter [103]. (**b**) Crystallinity vs. fiber diameter [103].

The mechanical properties of different nanofibers are shown in Table 2. According to Kancheva et al. (2015), the mechanical strength of various combinations of polylactic acid (PLA) and PCL mats was reported to be enhanced after thermal treatment at 60 °C. The melting of PCL enabled the sealing of the fibers, thus enhancing the mechanical properties of mats [157]. Wang et al. (2004) reported an increase in modulus of silk/PEO fibers from as-spun to methanol-treated and to water-extracted fibers [158]. The mechanical properties of single fibers were characterized by AFM nanoindentation. Tan et al. (2005) studied tensile property, using an approach that uses an atomic force microscope tip to stretch a single electrospun PEO nanofiber. The elastic modulus of PEO nanofiber was found to be 45 MPa [159]. Lin et al. (2012) demonstrated characterization of mechanical properties of ultra-thin electrospun polymer fibers. Electrospun techophilic, tecoflex, nylon 6, PVP, and PEOX fibers were captured directly on the testing device, stretched at controlled rates, and deflected with forces created by different velocities of streams of air [160].

Table 2. Mechanical properties of nanofibers.

Nanofibers	Tensile Strength	Tensile Modulus	Characteristics
PAN CNF	5.4 GPa	287 GPa	Hot drawn and carbonized at 1100 °C, 400 nm in diameter [139]
	7.3 GPa	262 GPa	Carbonized at 800 °C, 108 nm in diameter [104]
PCL	66 MPa	340 MPa	400 nm in diameter [102]
PVP	2.30 MPa [156] 7 MPa [160]	- 500 MPa [160]	300 nm in diameter [156] 800 nm in diameter [160]
PEO	- 45 MPa [159]	0.75 GPa [158] 22 MPa [159]	200 nm in diameter [158] 700 nm in diameter [159]
Nylon 6	900 MPa	304 MPa	800 nm in diameter [160]

PCL, polycaprolactone; PVP, polyvinyl pyrrolidone; PEO, polyethylene oxide.

3.2. Dielectric Properties

In this section, the dielectric properties of PAN fibers, carbon nanofillers, and other electrospun fibers and different nanomaterials are discussed. According to classical theory, the dielectric constant (k) is defined as the ratio of the permittivity (E) of a substance to the permittivity of free space (E_0). Values of k are always greater than or equal to 1. For most polymers, k values are in the range of 2 to 10.

3.2.1. PAN Nanofibers and Carbon Fillers

According to Li et al. (2010), the dielectric property depends on porosity and density [2]. The Figure 15 shows the dependence of density and porosity of PAN nanofiber membranes at the frequency of 1 MHz. The dielectric constant gradually increases with density ranging from 0.164 to 0.182 g/cm^3. Moreover, the dielectric decreases with increasing porosity from 84.4% to 86.1%. The apparent porosity can be found by Equation (6). Figure 16 shows the dielectric constant of PAN at 8 wt.% in the radio frequency range.

$$P\,(\%) = (1 - \frac{\rho_M}{\rho_P}) * 100 \tag{6}$$

where P is the porosity, ρ_M is the membrane density, and ρ_P is the polymer density.

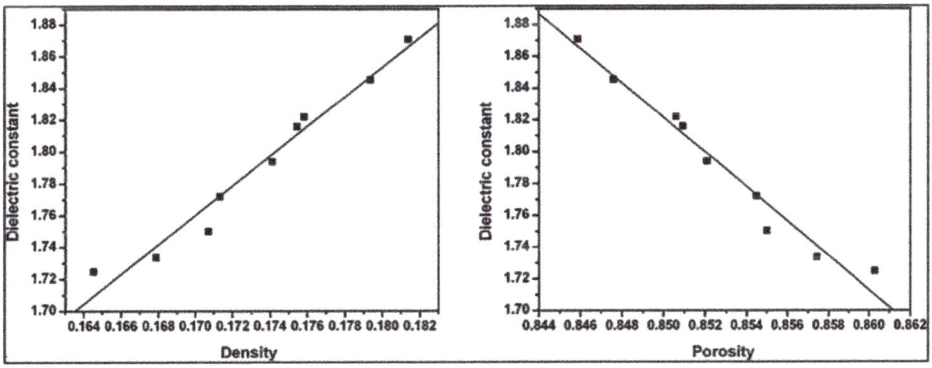

Figure 15. Dependence of density and porosity of PAN nanofiber membranes at frequency of 1 MHz [2].

Khan et al. (2014) reported the dielectric constant of PAN and PMMA as a function of graphene nano flakes as in Figure 17. The physical properties, including the dielectric constant, were significantly increased with graphene concentrations [161].

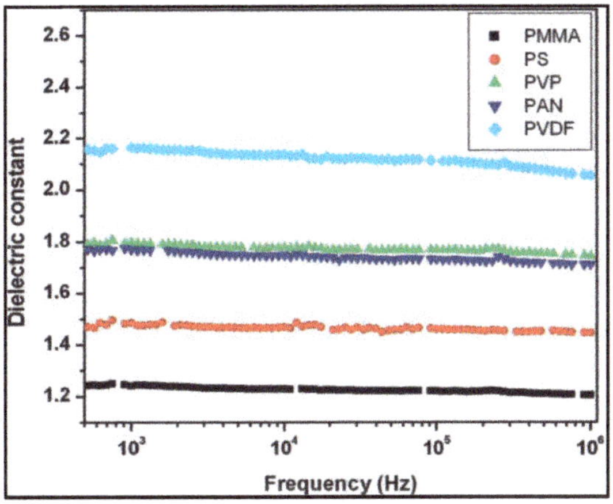

Figure 16. Dielectric constant of PAN at 8 wt.% in the radio frequency range [2].

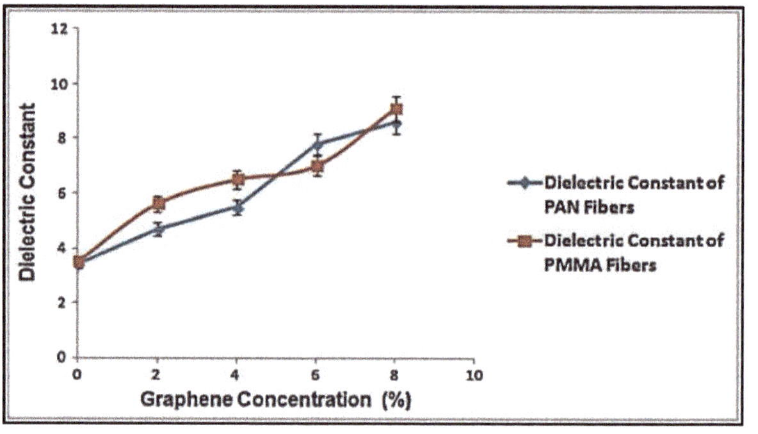

Figure 17. Dielectric constant of PAN as a function of graphene nano flakes concentrations [161].

According to Im et al. (2011), the increase in dielectric property of permittivity increases the electromagnetic interference shielding effectiveness (EMI-SE), as shown in Figure 18a,b. The permittivity of carbon fibers with different proportions of ZrO_2 is shown in Figure 18a, and the EMI-SE of the same proportions is shown in Figure 18b. This relationship is more evident below 2000 MHz [162].

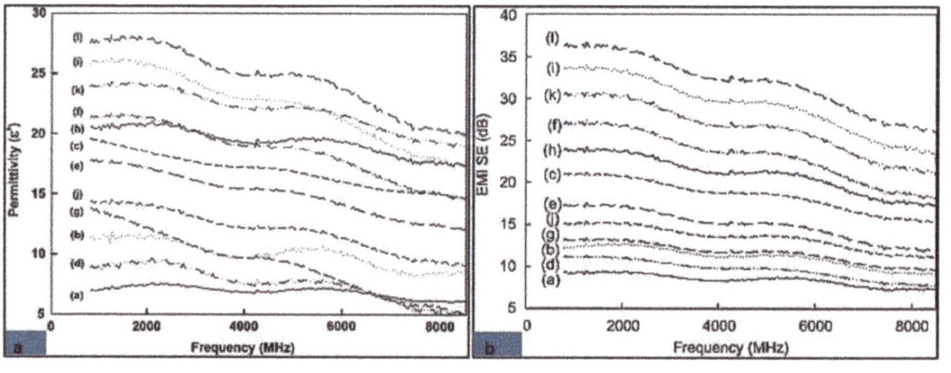

Figure 18. (**a**) Permittivity (ε′) and (**b**) electromagnetic interference shielding effectiveness (EMI-SE) of carbon fibers with different proportions of ZrO_2 [162].

Among many conventional carbon fillers, CNTs have been preferred for high dielectric constant. Bhattacharya (2016) gave a detailed review of processing of CNTs as potential nanofillers to form nanocomposites [163]. CNFs have been widely used in electrochemical energy storage devices as reviewed by Zhang et al. (2016) [12]. Electrospun polymer/CNF or CNT fibers have been used in energy storage devices. The comparison between the CNF and CNT is shown in Table 3.

Table 3. Comparison of carbon nanotube (CNT) and CNF [12].

Allotropes of Carbon	Specific Gravity (gcm^{-3})	Electric Conductivity (Scm^{-1})	Thermal Conductivity ($Wm^{-1}K^{-1}$)
CNF	1.5–2.0	10^{-7}–10^3	5–1600
CNT	0.8–1.8	10^2–10^6	2000–6000

3.2.2. Other Electrospun Fibers and Nanoparticles

The dielectric properties of different nanofibers are shown in Table 4. Lee et al. (2003) studied that dielectric constant strongly depends on solvent content and diameter of PCL electrospun fibers. As the solvent content increases, the fiber diameter decreases and dielectric constant increases [164]. Wei et al. (2014) reported the dielectric characterization of annealed electrospun $BaTiO_3$ fibers. Crystallized $BaTiO_3$ nanofibers showed better dielectric permittivity [165]. Electrospun PVDF fibers have higher β-crystalline content which enhances the piezoelectric property and its energy-harvesting application. Jabbarnia et al. (2016) reported various electrospun PVDF/PVP mats fabricated with different percentages of carbon black nanoparticles for the applications such as supercapacitor separators and other energy storage devices. The dielectric constant values were increased with the carbon black loading [166]. Lee et al. (2016) studied the effect of Fe and Co mixed with PVP. The analysis of FE-SEM images of electrospun products obtained by using solutions with and without citric acids was carried out. The composite showed excellent electromagnetic (EM) wave absorption properties where the power loss of the FeCo nanofibers increased to 20 GHz [167]. EM waves with frequencies in the microwave range of 12 to 18 GHz are widely used in wireless communication networks, radar systems, military aircraft, and satellite communication devices. Wang et al. (2011) studied that an increase in dielectric constant was achieved in a combination of high aspect ratio barium titanate ($BaTiO_3$) and graphene platelets in a silicon rubber matrix compared to their spherical counterparts. Higher volume fractions of ferroelectric particles lead to increased dielectric constant but also to lower mechanical properties. Composites with high aspect ratio fillers at lower loading exhibit higher dielectric constant [168]. Issa et al. (2017) reported an increase in permittivity by the addition of silver nanoparticles (AgNP) in PVDF. This is due to the interfacial polarization associated with entrapment of free charges generated at the interfaces between the AgNPs and PVDF [4]. Anita and Natarajan (2015) studied the potential of ZnO nanopowders with PVA matrix for the use of UV shielding [169].

Table 4. Dielectric properties of different nanoparticles.

Pure Polymers	Dielectric Constant	Property
PAN	~3.5	Physical properties can increase with graphene [161]
PMMA	~3.5	Physical properties can increase with graphene [161]
PVDF	~11	Physical properties can increase with addition of AgNP [4]
PVP	-	EMI increases with addition of FeCo (ε'~EMI) [167]
PVA	-	Uniform distribution of ZnO increases EMI (ε'~EMI) [169]
PU	-	EMI increases with PEDOT (ε'~EMI) [170]
PCL	~10	Dielectric increases with DMF concentration [164]

PMMA, polymethyl methacrylate; PVDF, polyvinylidene fluoride; PEDOT, conductive poly(3,4-ethylenedioxythiophene); DMF, dimethylformamide; PU, polyurethane.

The in situ sol–gel method (ISM) and direct deposition method (DDM) have been discussed to analyze the effect of UV shielding. The ISM–PVA/ZnO composite showed better UV absorption in optical transmission measurements due to the uniform dispersion of the ZnO in the fibrous matrix. Kim et al. (2016) analyzed the EMI-SE of multiwalled carbon nanotube (MWCNT) reinforced polyurethane (PU) in the DMF with tetrahydrofuran solvents and coated with conductive poly(3,4-ethylenedioxythiophene) (PEDOT) [170]. The EMI-SE from a network analyzer shows 25 dB at the frequency range of 50 MHz–10 GHz.

4. Applications of Aligned Fibers

The authors of this paper, Isaac et al. (2017), observed improvements in the mechanical strength and dielectric strength with increase in degree of alignment of fibers [171]. Aligned fibers are greatly beneficial when they are used in applications including field effect transistors, gas and optical sensors, fiber reinforced composite materials, and tissue engineering [172,173]. Bashur (2009) discussed the application of aligned fibers in the field of tissue engineering [174]. Moreover, Lawrence and Liu (2006) and Katti et al. (2004) stated that there are other applications found in the variety of areas if the fibers are in aligned form [175,176]. This section discusses mechanical and dielectric applications of aligned electrospun fibers.

4.1. Influence of Aligned Fibers on Mechanical Properties of Nanofiber Mats

Hou et al. (2005) showed that well-aligned, multi-walled carbon nanotubes (MWCNT) can improve the mechanical properties of a PAN-based nanofiber mat [177]. Kannan et al. (2007) demonstrated that electrospun polymer/CNT leads to nanocomposite fibers with embedded CNTs orienting parallel to the nanofiber axis [178]. Moreover, alignment of CNTs in the fiber direction can improve thermal conductivity [179]. Dhakate et al. (2016) reported that semi-aligned electrospun carbon nanofiber composites show excellent bending strength and interlaminar shear strength [180]. High-performance aramid copolymer fibers underwent four treatment factors. Among them, the degree of stretching after coagulation resulted in high degree of molecular orientation. The increased tensile strength of aramid fibers improved the cut resistance of aramid fibers, and therefore can be used in cut protection [181]. Ultra-high molecular weight polyethylene fibers with the tensile strength of 1.5 GPa were successfully prepared and structure and tensile property were studied. The increase in draw ratio improved the crystallinity of ultra high molecular weight polyethylene fibers. The molecular orientation degree increased, and tensile property also improved [182]. Increasing the draw-ratio resulted in an increased molecular orientation, Young's modulus, and tensile strength of poly(amide-block-aramid) fibers comprised of alternating rigid aramid blocks of poly(p-phenylene terephthalamide) (PPTA) and flexible blocks of polyamide 6,6. Heat treatment at 300 °C of the fibers resulted in an increase of Young's modulus and minor increase of strength [183].

Aligned micro scale fibers (7μm diameter) have application in composite reinforcement. There is increased need to manufacture complex composites for light weight applications. Carbon/epoxy composites have greater application in aircraft, sports cars, and space crafts because of a better strength to weight ratio than that of metals like aluminum alloys. They are thermally stable because of the lower coefficients of thermal expansion properties of carbon fibers. Yu et al. (2014) showed that short carbon fiber composites can be used in places where complex shapes and ductile properties are required [184]. Short carbon fibers, with an aspect ratio of 400, resulted in composites with a tensile modulus of 119 GPa and strength of 1211 GPa.

Compton and Lewis (2014) reported that cellular composites with controlled alignment of multi-scale and high aspect ratio fibers can result in reinforcement of hierarchical structures [185]. They demonstrated the first 3D printed cellular composites composed of oriented fiber-filled epoxy with exceptional mechanical properties. Malek et al. (2017) developed a new carbon fiber reinforced epoxy for 3D printing which resulted in printing materials with longitudinal Young's modulus up to 57 GPa [186].

4.2. Influence of Aligned Fibers on Dielectric Properties of Nanofiber Mats

Ning et al. (2014) showed that aligned MWCNT/PVA has high dielectric constant, low dielectric loss, high breakdown strength, and high energy density. These properties contribute in applications such as artificial muscles, energy storage, flexible electronics, and sensors [187]. Aligned MWCNT/PVA composite films were prepared by using electrospinning in situ film-forming technique. Additionally, Liu et al. (2012) confirmed the tailoring of dielectric property by controlling the alignment of CNTs [187]. Ma et al. (2012) reported that aligned PVDF had better molecular orientation than its random fiber counterparts. This is because of the smaller diameter of the aligned fibers. These nanofibers have applications in the field of sensors and actuators [188]. Agarwal et al. (2009) reported that aligned fibers have applications in nanofluidics, superhydrophobic patterning, nanoelectronics, and nanophotonic circuits [189]. Edmondson et al. (2012) demonstrated the significance of fiber alignment in improving the piezoelectric property, using centrifugal electrospinning. PVDF and PEO have piezo-, pyro-, and ferro-electric properties, and these aligned fibers can provide for applications in actuators, transistors, textiles, and composites [190]. P. Kumar et al. (2017) showed that aligned graphene films improved EMI shielding. Electromagnetic (EM) waves cause interference or device malfunction and also can cause harm to human bodies [191]. Song et al. (2013) observed that aligned carbon-based fillers enhanced EMI shielding. The alignment produced anisotropic characteristics that achieve enhancement in absorption and reflection performance [192].

5. Electrospinning System for Dielectric and Mechanical Property Studies

The authors of this paper, Isaac et al. (2017) studied the effect of electrospun fiber alignment on mechanical and dielectric properties, using a setup designed and drawn in 3D modeling software, as shown in Figure 19. The electrode, mandrel holding sheet, acrylonitrile butadiene styrene (ABS) sheet for adjusting the distance between needle tip and mandrel, and the sliding front door are shown above. The final physical setup of electrospinning device is shown in Figure 20.

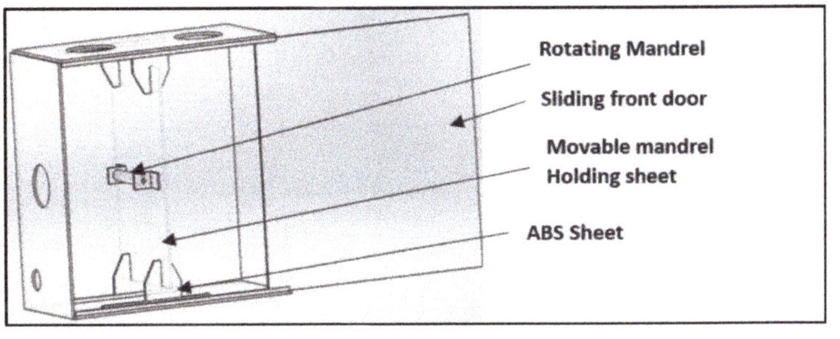

Figure 19. Enclosure, mandrel in the 3D model [171].

The whole apparatus is placed under a fume hood, for safety purposes, so that any toxic solvent escapes through the fume hood. Using this setup, electrospun fiber mats were fabricated, fiber morphology was analyzed, and tensile and dielectric properties were characterized.

Figure 20. The final electrospinning setup [171].

5.1. Material Builds

PAN/DMF solution was prepared with 8 wt.% concentration. The electrospinning device ran for ten minutes in order to form sufficient fibers on the aluminum foil wrapped on the mandrel. Figure 21a,b shows the fiber mat formed on the mandrel and the fiber mat unfolded from the mandrel.

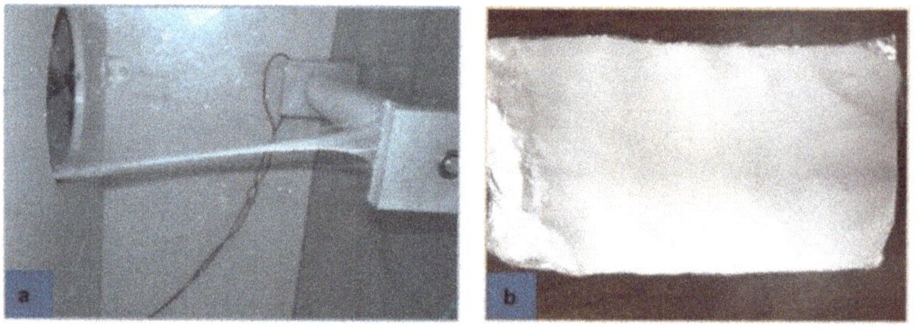

Figure 21. (**a**) Fiber mat is being formed on mandrel. (**b**) Fiber mat on an aluminum foil [171].

5.2. Fiber Morphological Analysis

The electrospun fiber mat formed was examined under a NanoSEM 230 SEM microscope to measure fiber diameter. As shown in Figure 22a,b, the fiber diameter decreases as the needle size decreases from 18G to 22G. The feed rate was 0.3 mL/h. The 18G needle produced fibers with average diameter of 1 μm, whereas the 22G needle produced fibers in the range of 300 to 900 nm diameters with an average diameter of 600 nm. The smaller the fiber diameter, the better the mechanical properties.

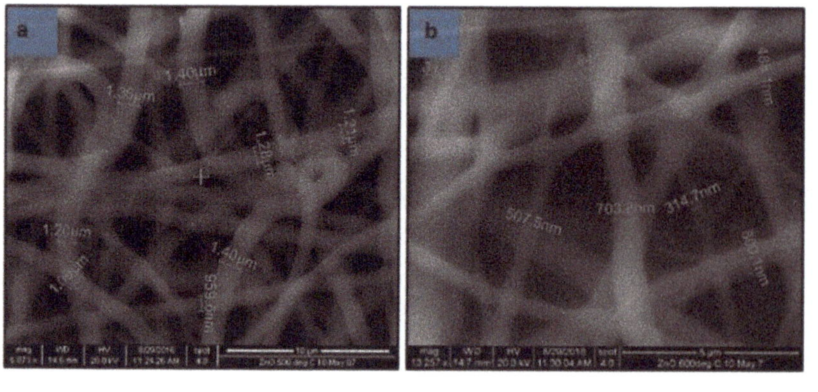

Figure 22. (**a**) Fiber mats using 18G needle magnified at 10 μm and (**b**) using 22G needle magnified at 5 μm [171].

5.3. Tensile and Dielectric Test Results of Electrospun Mats

PAN precursor with DMF solution at 8 wt.% concentration, using a 22G needle, is used for electrospinning. More than thirteen experimental runs were carried out to determine suitable process parameter settings. The three best samples were taken, to analyze the degree of fiber alignment. The tensile and dielectric test results for the three samples are given below in Figure 23a–c. The SEM images are shown in Figure 24.

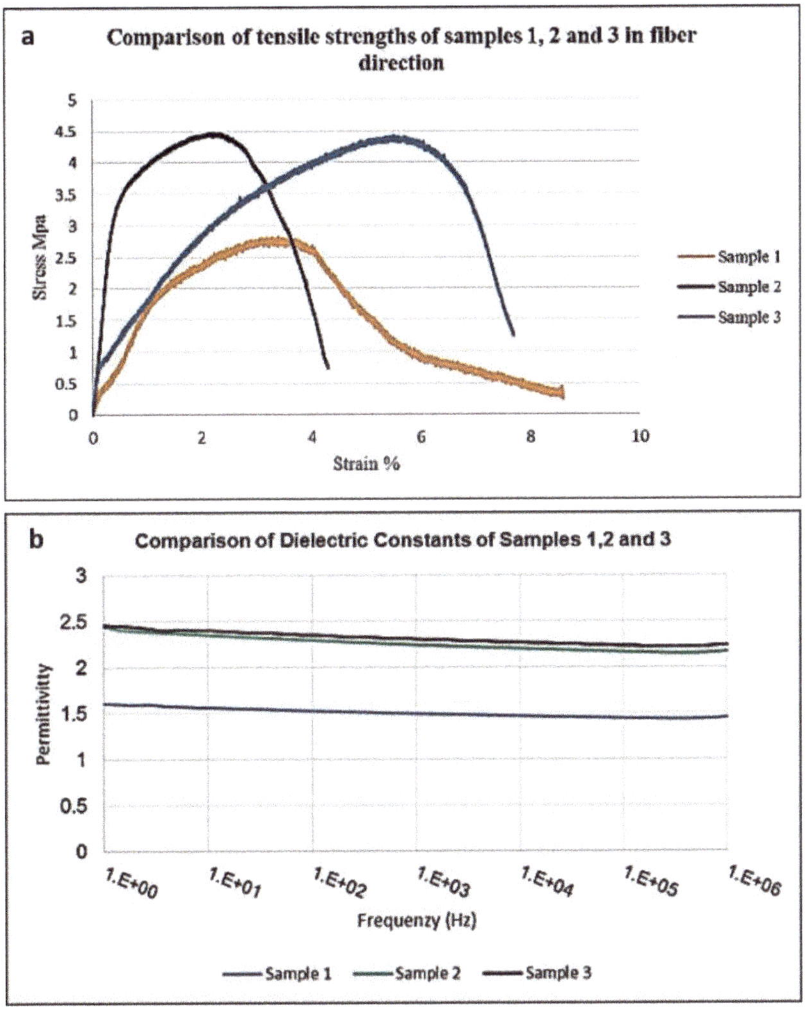

c Properties	Sample 1	Sample 2	Sample 3
Tensile Strength	2.82MPa	4.47MPa	4.43MPa
Dielectric Constant	1.65 @0.1Hz	2.52 @0.1Hz	2.56 @0.1Hz
SEM	Beads	No beads, but fibers broken	No beads
Diameter Range	723nm-1.7µm	822nm-1.2µm	620-760nm

Figure 23. (**a**) Tensile strength of best three samples [171]. (**b**) Dielectric strength of best three samples [171]. (**c**) Tensile and dielectric properties of three samples [171].

Figure 24. SEM images at 20 and 5 μm of Samples 1, 2, and 3 [171].

The SEM image of Sample 1 shows loosely packed nanofibers. Samples 2 and 3 have more closely packed fibers and better alignment of fibers, and Sample 3 has the highest densely packed fibers. These images are viewed at 20 μm magnification. Tensile strengths of Samples 2 and 3 are better than that of Sample 1, since they are better aligned and closely packed. As evident from Figure 23a, Samples 2 and 3 show higher tensile strength, and Sample 1 has the least. Better alignment combined with dense fibers contributed to the better tensile strengths of Samples 2 and 3. Sample 2 shows a tensile strength of 4.47 MPa. As shown below, in Figure 23b, Sample 3 shows the highest dielectric constant. The dielectric constant for Sample 3 is 2.56 Hz at 0.1 Hz. The best parameters are chosen by comparing the three given samples for their SEM images, tensile strengths, and dielectric properties as shown in Figure 23c. Sample 3 gives consistent and uniform fibers. The best parameters at 8 wt.% correspond to Sample 3. The Sample 3 has uniform fiber distribution with the lowest range in diameters. Moreover, the fibers are formed without beads. The tensile strength and dielectric constant of Sample 3 is found to be 4.43 MPa and 2.56, respectively.

6. Future Work

While a few studies have methodically experimented with system parameter optimization for limited sets of parameters, none has studied the nonlinear effects of two or more factors with three levels per factor for obtaining the mechanical and dielectric responses. Consequently, parameter interactions and nonlinearities have were studied, in-depth, to optimize alignment for mechanical and dielectric properties. Very few studies have been conducted on the nanofiber orientation variations along the fiber direction, using a rotating mandrel and its effect on mechanical and dielectric properties. The variability of data in both dielectric and tensile tests needs to be quantitatively addressed, and therefore a modified improved system should be developed. The improved system will lead to consistencies in the behavior of nanofiber specimens and acceptable standard deviations in variability that indicate meaningful parameter levels and trends. Future work will focus on system improvements and methodical experimentation (Design of Experiments), to optimize system parameters for improved mechanical and dielectric responses and application of PAN nanofiber mat materials in advanced mechanical and energy applications. Future research work will determine the nature of the trade-off between mechanical strength and dielectric properties—whether it is linear or non-linear, synergistic, or detrimental.

7. Conclusions

This work has provided a detailed review and analysis of methods for alignment of electrospun fibers and their resulting mechanical and dielectric properties. A key consideration is the molecular orientation of fibers along the fiber direction, as it is important for improvement in mechanical properties. Design configuration options for electrospinning apparatuses were surveyed in order to analyze the ability of each to improve fiber alignment. Given the mechanical and dielectric performance improvement possible with increasing fiber alignment, the need to improve electrospinning apparatus capabilities for fiber alignment is paramount and worthy of further study and experimentation.

Author Contributions: B.I., R.M.T. and K.R. discussed the contribution of electrospinning in various fields and outlined the manuscript with the focus on mechanical and dielectric properties. B.I wrote the manuscript, and R.M.T. and K.R. verified and edited it. All authors have read and agreed to the published version of the manuscript.

Funding: There is no funding for this project.

Acknowledgments: Blesson Isaac would like to thank Robert V. Fox, supervisor at Idaho National Laboratory for his guidance and The University of Texas at Arlington Research Institute for supplying the materials for experiments.

Conflicts of Interest: The authors declare that they have no conflict of interest.

References

1. Wan, L.Y.; Wang, H.; Gao, W.; Ko, F. An analysis of the tensile properties of nanofiber mats. *Polymer* **2015**, *73*, 62–67. [CrossRef]
2. Li, Y.; Lu, X.; Liu, X.; Zhang, C.; Li, X.; Zhang, W.; Wang, C. Ultra-low dielectric performance of polymer electrospun nanofiber mats. *Appl. Phys. A* **2010**, *100*, 207–212. [CrossRef]
3. Liu, L.; Lv, F.; Li, P.; Ding, L.; Tong, W.; Chu, P.K.; Zhang, Y. Preparation of ultra-low dielectric constant silica/polyimide nanofiber membranes by electrospinning. *Compos. Part A Appl. Sci. Manuf.* **2016**, *84*, 292–298. [CrossRef]
4. Issa, A.A.; Al-Maadeed, M.A.; Luyt, A.S.; Ponnamma, D.; Hassan, M.K. Physico-mechanical, dielectric, and piezoelectric properties of PVDF electrospun mats containing silver nanoparticles. *C J. Carbon Res.* **2017**, *3*, 30. [CrossRef]
5. Dwivedi, S.; Sakamoto, S.; Kato, S.; Mitsumata, T.; Kaneko, T. Effects of biopolyimide molecular design on their silica hybrids thermo-mechanical, optical and electrical properties. *RSC Adv.* **2018**, *8*, 14009–14016. [CrossRef]

6. Bhardwaj, N.; Kundu, S.C. Electrospinning: A fascinating fiber fabrication technique. *Biotechnol. Adv.* **2010**, *28*, 325–347. [CrossRef]
7. Cooley, J.F. Improved Methods of and Apparatus for Electrically Separating the Relatively Volatile Liquid Component from the Component of Relatively Fixed Substances of Composite Fluids. *United Kingd. Pat.* **1900**, *6385*, 19.
8. Cooley, J.F. Apparatus for Electrically Dispersing Fluids. U.S. Patents 6,926,31A, 4 February 1902.
9. Morton, W.J. Method of Dispersing Fluids. U.S. Patents 7,056,91A, 29 July 1902.
10. Anton, F. Process and Apparatus for Preparing Artificial Threads. U.S. Patents 19,755,04A, 2 October 1934.
11. Anton, F. Method and Apparatus for Spinning. U.S. Patents 21,609,62A, 6 June 1939.
12. Zhang, B.; Kang, F.; Tarascon, J.-M.; Kim, J.-K. Recent advances in electrospun carbon nanofibers and their application in electrochemical energy storage. *Prog. Mater. Sci.* **2016**, *76*, 319–380. [CrossRef]
13. Nataraj, S.; Yang, K.; Aminabhavi, T. Polyacrylonitrile-based nanofibers—A state-of-the-art review. *Prog. Polym. Sci.* **2012**, *37*, 487–513. [CrossRef]
14. Luo, C.; Stoyanov, S.D.; Stride, E.; Pelan, E.; Edirisinghe, M. Electrospinning versus fibre production methods: From specifics to technological convergence. *Chem. Soc. Rev.* **2012**, *41*, 4708–4735. [CrossRef]
15. Zhang, C.; Feng, F.; Zhang, H. Emulsion electrospinning: Fundamentals, food applications and prospects. *Trends Food Sci. Technol.* **2018**, *80*, 175–186. [CrossRef]
16. Taylor, G.I. Electrically driven jets. *Proc. R. Soc. Lond. A Math. Phys. Sci.* **1969**, *313*, 453–475.
17. Huang, Z.-M.; Zhang, Y.Z.; Kotaki, M.; Ramakrishna, S. A review on polymer nanofibers by electrospinning and their applications in nanocomposites. *Compos. Sci. Technol.* **2003**, *63*, 2223–2253. [CrossRef]
18. Afshari, M. 4.3.1.3 Effect of Electric Charge. In *Electrospun Nanofibers*; Elsevier: Amsterdam, The Netherlands, 2017.
19. Reneker, D.H.; Yarin, A.L. Electrospinning jets and polymer nanofibers. *Polymer* **2008**, *49*, 2387–2425. [CrossRef]
20. Gibson, P.; Schreuder-Gibson, H.; Rivin, D. Transport properties of porous membranes based on electrospun nanofibers. *Colloids Surf. A Physicochem. Eng. Asp.* **2001**, *187*, 469–481. [CrossRef]
21. Bruce, P.G.; Freunberger, S.A.; Hardwick, L.J.; Tarascon, J.-M. Li–O2 and Li–S batteries with high energy storage. *Nat. Mater.* **2012**, *11*, 19–29. [CrossRef]
22. Goodenough, J.B. Electrochemical energy storage in a sustainable modern society. *Energy Environ. Sci.* **2014**, *7*, 14–18. [CrossRef]
23. Peng, S.; Jin, G.; Li, L.; Li, K.; Srinivasan, M.; Ramakrishna, S.; Chen, J. Multi-functional electrospun nanofibres for advances in tissue regeneration, energy conversion & storage, and water treatment. *Chem. Soc. Rev.* **2016**, *45*, 1225–1241.
24. Khil, M.S.; Cha, D.I.; Kim, H.Y.; Kim, I.S.; Bhattarai, N. Electrospun nanofibrous polyurethane membrane as wound dressing. *J. Biomed. Mater. Res. Part B* **2003**, *67*, 675–679. [CrossRef]
25. Mao, X.; Hatton, T.A.; Rutledge, G.C. A review of electrospun carbon fibers as electrode materials for energy storage. *Curr. Org. Chem.* **2013**, *17*, 1390–1401. [CrossRef]
26. Mei, Y.; Yao, C.; Fan, K.; Li, X. Surface modification of polyacrylonitrile nanofibrous membranes with superior antibacterial and easy-cleaning properties through hydrophilic flexible spacers. *J. Membr. Sci.* **2012**, *417*, 20–27. [CrossRef]
27. Jiang, S.; Chen, Y.; Duan, G.; Mei, C.; Greiner, A.; Agarwal, S. Electrospun nanofiber reinforced composites: A review. *Polym. Chem.* **2018**, *9*, 2685–2720. [CrossRef]
28. Bergshoef, M.M.; Vancso, G.J. Transparent nanocomposites with ultrathin, electrospun nylon-4, 6 fiber reinforcement. *Adv. Mater.* **1999**, *11*, 1362–1365. [CrossRef]
29. Ahmed, F.E.; Lalia, B.S.; Hashaikeh, R. A review on electrospinning for membrane fabrication: Challenges and applications. *Desalination* **2015**, *356*, 15–30. [CrossRef]
30. Musiari, F.; Pirondi, A.; Moroni, F.; Giuliese, G.; Belcari, J.; Zucchelli, A.; Brugo, T.; Minak, G.; Ragazzini, C. Feasibility study of adhesive bonding reinforcement by electrospun nanofibers. *Procedia Struct. Integr.* **2016**, *2*, 112–119. [CrossRef]
31. Mirjalili, M.; Zohoori, S. Review for application of electrospinning and electrospun nanofibers technology in textile industry. *J. Nanostruct. Chem.* **2016**, *6*, 207–213. [CrossRef]
32. Sekiya, N.; Ichioka, S.; Terada, D.; Tsuchiya, S.; Kobayashi, H. Efficacy of a poly glycolic acid (PGA)/collagen composite nanofibre scaffold on cell migration and neovascularisation in vivo skin defect model. *J. Plast. Surg. Hand Surg.* **2013**, *47*, 498–502. [CrossRef]
33. Kumbar, S.G.; Nukavarapu, S.P.; James, R.; Nair, L.S.; Laurencin, C.T. Electrospun poly(lactic acid-co-glycolic acid) scaffolds for skin tissue engineering. *Biomaterials* **2008**, *29*, 4100–4107. [CrossRef]
34. Jin, G.; Prabhakaran, M.P.; Ramakrishna, S. Stem cell differentiation to epidermal lineages on electrospun nanofibrous substrates for skin tissue engineering. *Acta Biomater.* **2011**, *7*, 3113–3122. [CrossRef]
35. Mukhatyar, V.J.; Salmerón-Sánchez, M.; Rudra, S.; Mukhopadaya, S.; Barker, T.H.; García, A.J.; Bellamkonda, R.V. Role of fibronectin in topographical guidance of neurite extension on electrospun fibers. *Biomaterials* **2011**, *32*, 3958–3968. [CrossRef]
36. Lim, S.H.; Liu, X.Y.; Song, H.; Yarema, K.J.; Mao, H.-Q. The effect of nanofiber-guided cell alignment on the preferential differentiation of neural stem cells. *Biomaterials* **2010**, *31*, 9031–9039. [CrossRef] [PubMed]
37. Xie, J.; Liu, W.; MacEwan, M.R.; Bridgman, P.C.; Xia, Y. Neurite Outgrowth on Electrospun Nanofibers with Uniaxial Alignment: The Effects of Fiber Density, Surface Coating, and Supporting Substrate. *ACS Nano* **2014**, *8*, 1878–1885. [CrossRef] [PubMed]
38. Zong, X.; Bien, H.; Chung, C.Y.; Yin, L.; Fang, D.; Hsiao, B.S.; Chu, B.; Entcheva, E. Electrospun fine-textured scaffolds for heart tissue constructs. *Biomaterials* **2005**, *26*, 5330–5338. [CrossRef] [PubMed]

39. Kharaziha, M.; Nikkhah, M.; Shin, S.R.; Annabi, N.; Masoumi, N.; Gaharwar, A.K.; Camci-Unal, G.; Khademhosseini, A. PGS:Gelatin nanofibrous scaffolds with tunable mechanical and structural properties for engineering cardiac tissues. *Biomaterials* **2013**, *34*, 6355–6366. [CrossRef] [PubMed]
40. Hussain, A.; Collins, G.; Yip, D.; Cho, C.H. Functional 3-D cardiac co-culture model using bioactive chitosan nanofiber scaffolds. *Biotechnol. Bioeng.* **2013**, *110*, 637–647. [CrossRef] [PubMed]
41. Nerurkar, N.L.; Baker, B.M.; Sen, S.; Wible, E.E.; Elliott, D.M.; Mauck, R.L. Nanofibrous biologic laminates replicate the form and function of the annulus fibrosus. *Nat. Mater.* **2009**, *8*, 986–992. [CrossRef]
42. Surrao, D.C.; Waldman, S.D.; Amsden, B.G. Biomimetic poly(lactide) based fibrous scaffolds for ligament tissue engineering. *Acta Biomater.* **2012**, *8*, 3997–4006. [CrossRef]
43. Coburn, J.M.; Gibson, M.; Monagle, S.; Patterson, Z.; Elisseeff, J.H. Bioinspired nanofibers support chondrogenesis for articular cartilage repair. *Proc. Natl. Acad. Sci. USA* **2012**, *109*, 10012–10017. [CrossRef]
44. Kumar, P.S.; Sundaramurthy, J.; Sundarrajan, S.; Babu, V.J.; Singh, G.; Allakhverdiev, S.I.; Ramakrishna, S. Hierarchical electrospun nanofibers for energy harvesting, production and environmental remediation. *Energy Environ. Sci.* **2014**, *7*, 3192–3222. [CrossRef]
45. Krishnamoorthy, T.; Thavasi, V.; Ramakrishna, S. A first report on the fabrication of vertically aligned anatase TiO 2 nanowires by electrospinning: Preferred architecture for nanostructured solar cells. *Energy Environ. Sci.* **2011**, *4*, 2807–2812. [CrossRef]
46. Kumar, E.N.; Jose, R.; Archana, P.; Vijila, C.; Yusoff, M.; Ramakrishna, S. High performance dye-sensitized solar cells with record open circuit voltage using tin oxide nanoflowers developed by electrospinning. *Energy Environ. Sci.* **2012**, *5*, 5401–5407. [CrossRef]
47. Batmunkh, M.; Macdonald, T.J.; Shearer, C.J.; Bat-Erdene, M.; Wang, Y.; Biggs, M.J.; Parkin, I.P.; Nann, T.; Shapter, J.G. Carbon nanotubes in TiO2 nanofiber photoelectrodes for high-performance perovskite solar cells. *Adv. Sci.* **2017**, *4*, 1600504. [CrossRef] [PubMed]
48. Dong, Z.; Kennedy, S.J.; Wu, Y. Electrospinning materials for energy-related applications and devices. *J. Power Sources* **2011**, *196*, 4886–4904. [CrossRef]
49. Liu, D.; Guo, Q.; Hou, H.; Niwa, O.; You, T. PdxCoy Nanoparticle/Carbon Nanofiber Composites with Enhanced Electrocatalytic Properties. *ACS Catal.* **2014**, *4*, 1825–1829. [CrossRef]
50. Dong, B.; Gwee, L.; Salas-de la Cruz, D.; Winey, K.I.; Elabd, Y.A. Super Proton Conductive High-Purity Nafion Nanofibers. *Nano Lett.* **2010**, *10*, 3785–3790. [CrossRef]
51. Kim, M.; Kwon, C.; Eom, K.; Kim, J.; Cho, E. Electrospun Nb-doped TiO_2 nanofiber support for Pt nanoparticles with high electrocatalytic activity and durability. *Sci. Rep.* **2017**, *7*, 44411. [CrossRef]
52. Cheng, F.; Liang, J.; Tao, Z.; Chen, J. Functional materials for rechargeable batteries. *Adv. Mater.* **2011**, *23*, 1695–1715. [CrossRef]
53. Xu, J.J.; Xu, D.; Wang, Z.L.; Wang, H.G.; Zhang, L.L.; Zhang, X.B. Synthesis of perovskite-based porous La0. 75Sr0. 25MnO3 nanotubes as a highly efficient electrocatalyst for rechargeable lithium-oxygen batteries. *Angew. Chem. Int. Ed.* **2013**, *52*, 3887–3890. [CrossRef]
54. Sun, Y.; Sills, R.B.; Hu, X.; Seh, Z.W.; Xiao, X.; Xu, H.; Luo, W.; Jin, H.; Xin, Y.; Li, T.; et al. A Bamboo-Inspired Nanostructure Design for Flexible, Foldable, and Twistable Energy Storage Devices. *Nano Lett.* **2015**, *15*, 3899–3906. [CrossRef]
55. Zhang, X.; Suresh Kumar, P.; Aravindan, V.; Liu, H.H.; Sundaramurthy, J.; Mhaisalkar, S.G.; Duong, H.M.; Ramakrishna, S.; Madhavi, S. Electrospun TiO_2—Graphene composite nanofibers as a highly durable insertion anode for lithium ion batteries. *J. Phys. Chem. C* **2012**, *116*, 14780–14788. [CrossRef]
56. Lee, C.-G.; Javed, H.; Zhang, D.; Kim, J.-H.; Westerhoff, P.; Li, Q.; Alvarez, P.J.J. Porous Electrospun Fibers Embedding TiO_2 for Adsorption and Photocatalytic Degradation of Water Pollutants. *Environ. Sci. Technol.* **2018**, *52*, 4285–4293. [CrossRef] [PubMed]
57. Liao, Y.; Loh, C.-H.; Tian, M.; Wang, R.; Fane, A.G. Progress in electrospun polymeric nanofibrous membranes for water treatment: Fabrication, modification and applications. *Prog. Polym. Sci.* **2018**, *77*, 69–94. [CrossRef]
58. Zhu, F.; Zheng, Y.-M.; Zhang, B.-G.; Dai, Y.-R. A critical review on the electrospun nanofibrous membranes for the adsorption of heavy metals in water treatment. *J. Hazard. Mater.* **2020**, 123608. [CrossRef] [PubMed]
59. Roongraung, K.; Chuangchote, S.; Laosiripojana, N.; Sagawa, T. Electrospun Ag-TiO_2 Nanofibers for Photocatalytic Glucose Conversion to High-Value Chemicals. *ACS Omega* **2020**, *5*, 5862–5872. [CrossRef] [PubMed]
60. Blanco, M.; Monteserín, C.; Angulo, A.; Pérez-Márquez, A.; Maudes, J.; Murillo, N.; Aranzabe, E.; Ruiz-Rubio, L.; Vilas, J.L. TiO_2-doped electrospun nanofibrous membrane for photocatalytic water treatment. *Polymers* **2019**, *11*, 747. [CrossRef] [PubMed]
61. Norouzi, M.; Fazeli, A.; Tavakoli, O. Phenol contaminated water treatment by photocatalytic degradation on electrospun Ag/TiO_2 nanofibers: Optimization by the response surface method. *J. Water Process Eng.* **2020**, *37*, 101489. [CrossRef]
62. Khin, M.M.; Nair, A.S.; Babu, V.J.; Murugan, R.; Ramakrishna, S. A review on nanomaterials for environmental remediation. *Energy Environ. Sci.* **2012**, *5*, 8075–8109. [CrossRef]
63. Das, R.; Ali, M.E.; Hamid, S.B.A.; Ramakrishna, S.; Chowdhury, Z.Z. Carbon nanotube membranes for water purification: A bright future in water desalination. *Desalination* **2014**, *336*, 97–109. [CrossRef]
64. Kaur, S.; Gopal, R.; Ng, W.J.; Ramakrishna, S.; Matsuura, T. Next-Generation Fibrous Media for Water Treatment. *Mrs Bull.* **2008**, *33*, 21–26. [CrossRef]
65. Teo, W.E.; Ramakrishna, S. A review on electrospinning design and nanofibre assemblies. *Nanotechnology* **2006**, *17*, R89. [CrossRef]
66. Pham, Q.P.; Sharma, U.; Mikos, A.G. Electrospinning of polymeric nanofibers for tissue engineering applications: A review. *Tissue Eng.* **2006**, *12*, 1197–1211. [CrossRef] [PubMed]

67. Shuakat, M.N.; Lin, T. Recent developments in electrospinning of nanofiber yarns. *J. Nanosci. Nanotechnol.* **2014**, *14*, 1389–1408. [CrossRef] [PubMed]
68. Shi, X.; Zhou, W.; Ma, D.; Ma, Q.; Bridges, D.; Ma, Y.; Hu, A. Electrospinning of nanofibers and their applications for energy devices. *J. Nanomater.* **2015**, *2015*. [CrossRef]
69. Shekh, M.I.; Patel, N.N.; Patel, K.P.; Patel, R.M.; Ray, A. Nano silver-embedded electrospun nanofiber of poly (4-chloro-3-methylphenyl methacrylate): Use as water sanitizer. *Environ. Sci. Pollut. Res.* **2017**, *24*, 5701–5716. [CrossRef] [PubMed]
70. Li, Y.; Huang, X.; Zeng, L.; Li, R.; Tian, H.; Fu, X.; Wang, Y.; Zhong, W.-H. A review of the electrical and mechanical properties of carbon nanofiller-reinforced polymer composites. *J. Mater. Sci.* **2019**, *54*, 1036–1076. [CrossRef]
71. Zafar, M.; Najeeb, S.; Khurshid, Z.; Vazirzadeh, M.; Zohaib, S.; Najeeb, B.; Sefat, F. Potential of electrospun nanofibers for biomedical and dental applications. *Materials* **2016**, *9*, 73. [CrossRef]
72. Zhao, Z.; Li, J.; Yuan, X.; Li, X.; Zhang, Y.; Sheng, J. Preparation and properties of electrospun poly (vinylidene fluoride) membranes. *J. Appl. Polym. Sci.* **2005**, *97*, 466–474. [CrossRef]
73. Zeng, J.; Haoqing, H.; Schaper, A.; Wendorff, J.H.; Greiner, A. Poly-L-lactide nanofibers by electrospinning–Influence of solution viscosity and electrical conductivity on fiber diameter and fiber morphology. *e-Polymers* **2003**, *3*. [CrossRef]
74. Gupta, P.; Elkins, C.; Long, T.E.; Wilkes, G.L. Electrospinning of linear homopolymers of poly (methyl methacrylate): Exploring relationships between fiber formation, viscosity, molecular weight and concentration in a good solvent. *Polymer* **2005**, *46*, 4799–4810. [CrossRef]
75. Koski, A.; Yim, K.; Shivkumar, S. Effect of molecular weight on fibrous PVA produced by electrospinning. *Mater. Lett.* **2004**, *58*, 493–497. [CrossRef]
76. Mit-uppatham, C.; Nithitanakul, M.; Supaphol, P. Ultrafine electrospun polyamide-6 fibers: Effect of solution conditions on morphology and average fiber diameter. *Macromol. Chem. Phys.* **2004**, *205*, 2327–2338. [CrossRef]
77. Fong, H.; Chun, I.; Reneker, D.H. Beaded nanofibers formed during electrospinning. *Polymer* **1999**, *40*, 4585–4592. [CrossRef]
78. Fang, J.; Wang, H.; Niu, H.; Lin, T.; Wang, X. Evolution of fiber morphology during electrospinning. *J. Appl. Polym. Sci.* **2010**, *118*, 2553–2561. [CrossRef]
79. Uyar, T.; Besenbacher, F. Electrospinning of uniform polystyrene fibers: The effect of solvent conductivity. *Polymer* **2008**, *49*, 5336–5343. [CrossRef]
80. Zuo, W.; Zhu, M.; Yang, W.; Yu, H.; Chen, Y.; Zhang, Y. Experimental study on relationship between jet instability and formation of beaded fibers during electrospinning. *Polym. Eng. Sci.* **2005**, *45*, 704–709. [CrossRef]
81. Demir, M.M.; Yilgor, I.; Yilgor, E.; Erman, B. Electrospinning of polyurethane fibers. *Polymer* **2002**, *43*, 3303–3309. [CrossRef]
82. Ki, C.S.; Baek, D.H.; Gang, K.D.; Lee, K.H.; Um, I.C.; Park, Y.H. Characterization of gelatin nanofiber prepared from gelatin–formic acid solution. *Polymer* **2005**, *46*, 5094–5102. [CrossRef]
83. Sill, T.J.; Von Recum, H.A. Electrospinning: Applications in drug delivery and tissue engineering. *Biomaterials* **2008**, *29*, 1989–2006. [CrossRef]
84. Lee, K.H.; Kim, H.Y.; Bang, H.J.; Jung, Y.H.; Lee, S.G. The change of bead morphology formed on electrospun polystyrene fibers. *Polymer* **2003**, *44*, 4029–4034. [CrossRef]
85. Casper, C.L.; Stephens, J.S.; Tassi, N.G.; Chase, D.B.; Rabolt, J.F. Controlling surface morphology of electrospun polystyrene fibers: Effect of humidity and molecular weight in the electrospinning process. *Macromolecules* **2004**, *37*, 573–578. [CrossRef]
86. Reneker, D.H.; Chun, I. Nanometre diameter fibres of polymer, produced by electrospinning. *Nanotechnology* **1996**, *7*, 216. [CrossRef]
87. De Vrieze, S.; Van Camp, T.; Nelvig, A.; Hagström, B.; Westbroek, P.; De Clerck, K. The effect of temperature and humidity on electrospinning. *J. Mater. Sci.* **2009**, *44*, 1357–1362. [CrossRef]
88. Gu, S.; Ren, J.; Vancso, G. Process optimization and empirical modeling for electrospun polyacrylonitrile (PAN) nanofiber precursor of carbon nanofibers. *Eur. Polym. J.* **2005**, *41*, 2559–2568. [CrossRef]
89. Senthil, T.; Anandhan, S. Fabrication of styrene–acrylonitrile random copolymer nanofiber membranes from N, N-dimethyl formamide by electrospinning. *J. Elastomers Plast.* **2015**, *47*, 327–346. [CrossRef]
90. Isaac, B.; Taylor, R.M.; Reifsnider, K. Anisotropic Characterizations of Electrospun PAN Nanofiber Mats Using Design of Experiments. *Nanomaterials* **2020**, *10*, 2273. [CrossRef]
91. Isaac, B.; Taylor, R.M. *Electrospinning Approach for the Improvement of Mechanical and Dielectric Properties of Anisotropic Nanofiber Mat by Using a Novel Fiber Alignment Technique*; The University of Texas at Arlington: Arlington, TX, USA, 2018.
92. Rafiei, S.; Maghsoodloo, S.; Noroozi, B.; Mottaghitalab, V.; Haghi, A. Mathematical modeling in electrospinning process of nanofibers: A detailed review. *Cellul. Chem. Technol.* **2013**, *47*, 323–338.
93. Ismail, N.; Maksoud, F.J.; Ghaddar, N.; Ghali, K.; Tehrani-Bagha, A. Simplified modeling of the electrospinning process from the stable jet region to the unstable region for predicting the final nanofiber diameter. *J. Appl. Polym. Sci.* **2016**, 133. [CrossRef]
94. Rafiei, S.; Maghsoodloo, S.; Saberi, M.; Lotfi, S.; Motaghitalab, V.; Noroozi, B.; Haghi, A. New horizons in modeling and simulation of electrospun nanofibers: A detailed review. *Cellul. Chem. Technol.* **2014**, *48*, 401–424.
95. Kowalewski, T.A.; Barral, S.; Kowalczyk, T. Modeling electrospinning of nanofibers. In Proceedings of the IUTAM Symposium on Modelling Nanomaterials and Nanosystems, Aalborg, Denmark, 19–22 May 2008; pp. 279–292.
96. Kowalewski, T.; Błoński, S.; Barral, S. Experiments and modelling of electrospinning process. *Bull. Pol. Acad. Sci. Tech. Sci.* **2005**, *53*, 385–394.

97. Ghaly, M. Fem of Electrospinning Compared to Inkjet Printing Model. Master's Thesis, New Jersey Institute of Technology, Newark, NJ, USA, 2014.
98. Yee, W.A.; Kotaki, M.; Liu, Y.; Lu, X. Morphology, polymorphism behavior and molecular orientation of electrospun poly(vinylidene fluoride) fibers. *Polymer* **2007**, *48*, 512–521. [CrossRef]
99. Fennessey, S.F.; Farris, R.J. Fabrication of aligned and molecularly oriented electrospun polyacrylonitrile nanofibers and the mechanical behavior of their twisted yarns. *Polymer* **2004**, *45*, 4217–4225. [CrossRef]
100. Isaac, B.; Vaagensmith, B.C.; Reeves, J.L. *Silica Nanofiber Mat for Thermal Insulator Using Electrospinning*; Idaho National Lab (INL): Idaho Falls, ID, USA, 2019.
101. Bahl, O.P.; Mathur, R.B.; Kundra, K.D. Structure of PAN fibres and its relationship to resulting carbon fibre properties. *Fibre Sci. Technol.* **1981**, *15*, 147–151. [CrossRef]
102. Chari, S.S.; Bahl, O.P.; Mathur, R.B. Characterisation of acrylic fibres used for making carbon fibres. *Fibre Sci. Technol.* **1981**, *15*, 153–160. [CrossRef]
103. Baji, A.; Mai, Y.-W.; Wong, S.-C.; Abtahi, M.; Chen, P. Electrospinning of polymer nanofibers: Effects on oriented morphology, structures and tensile properties. *Compos. Sci. Technol.* **2010**, *70*, 703–718. [CrossRef]
104. Beese, A.M.; Papkov, D.; Li, S.; Dzenis, Y.; Espinosa, H.D. In situ transmission electron microscope tensile testing reveals structure–property relationships in carbon nanofibers. *Carbon* **2013**, *60*, 246–253. [CrossRef]
105. Arshad, S.N.; Naraghi, M.; Chasiotis, I. Strong carbon nanofibers from electrospun polyacrylonitrile. *Carbon* **2011**, *49*, 1710–1719. [CrossRef]
106. Uyar, T.; Cianga, I.; Cianga, L.; Besenbacher, F.; Yagci, Y. Self-aligned and bundled electrospun fibers prepared from blends of polystyrene (PS) and poly (methyl methacrylate)(PMMA) with a hairy-rod polyphenylene copolymer. *Mater. Lett.* **2009**, *63*, 1638–1641. [CrossRef]
107. Alfaro De Prá, M.A.; Ribeiro-do-Valle, R.M.; Maraschin, M.; Veleirinho, B. Effect of collector design on the morphological properties of polycaprolactone electrospun fibers. *Mater. Lett.* **2017**, *193*, 154–157. [CrossRef]
108. Han, T.H.; Nirmala, R.; Kim, T.W.; Navamathavan, R.; Kim, H.Y.; Park, S.J. Highly aligned poly(vinylidene fluoride-co-hexafluoro propylene) nanofibers via electrospinning technique. *J. Nanosci. Nanotechnol.* **2016**, *16*, 595–600. [CrossRef]
109. Kanu, N.J.; Gupta, E.; Vates, U.K.; Singh, G.K. Electrospinning process parameters optimization for biofunctional curcumin/gelatin nanofibers. *Mater. Res. Express* **2020**, *7*, 035022. [CrossRef]
110. Munir, M.M.; Nuryantini, A.Y.; Iskandar; Suciati, T.; Khairurrijal, K. Mass Production of Stacked Styrofoam Nanofibers Using a Multinozzle and Drum Collector Electrospinning System. *Adv. Mater. Res.* **2014**, *896*, 20–23. [CrossRef]
111. Yee, W.A.; Nguyen, A.C.; Lee, P.S.; Kotaki, M.; Liu, Y.; Tan, B.T.; Mhaisalkar, S.; Lu, X. Stress-induced structural changes in electrospun polyvinylidene difluoride nanofibers collected using a modified rotating disk. *Polymer* **2008**, *49*, 4196–4203. [CrossRef]
112. El-hadi, A.M.; Al-Jabri, F.Y. Influence of electrospinning parameters on fiber diameter and mechanical properties of poly (3-hydroxybutyrate)(PHB) and polyanilines (PANI) blends. *Polymers* **2016**, *8*, 97. [CrossRef] [PubMed]
113. Kameoka, J.; Craighead, H. Fabrication of oriented polymeric nanofibers on planar surfaces by electrospinning. *Appl. Phys. Lett.* **2003**, *83*, 371–373. [CrossRef]
114. Yu, L.; Shao, Z.; Xu, L.; Wang, M. High throughput preparation of aligned nanofibers using an improved bubble-electrospinning. *Polymers* **2017**, *9*, 658. [CrossRef]
115. Lingaiah, S.; Shivakumar, K.N.; Sadler, R.; Sharpe, M. Electrospinning of nanofabrics. In Proceedings of the SAMPE '07: M and P—From Coast to Coast and Around the World, Baltimore, MD, USA, 3–7 June 2007. SAMPE Baltimore/Washington Chapter.
116. Theron, A.; Zussman, E.; Yarin, A. Electrostatic field-assisted alignment of electrospun nanofibres. *Nanotechnology* **2001**, *12*, 384. [CrossRef]
117. Karayeğen, G.; Koçum, İ.C.; Çökeliler Serdaroğlu, D.; Doğan, M. Aligned polyvinylpyrrolidone nanofibers with advanced electrospinning for biomedical applications. *Bio-Med Mater. Eng.* **2018**, *29*, 685–697. [CrossRef]
118. Liu, H.Y.; Xu, L.; Tang, X.P.; Si, N. Fabrication of aligned PAN nanofiber by electrospinning with parallel electrode. *Adv. Mater. Res.* **2014**, *905*, 19–22. [CrossRef]
119. Secasanu, V.P.; Giardina, C.K.; Wang, Y. A novel electrospinning target to improve the yield of uniaxially aligned fibers. *Biotechnol. Prog.* **2009**, *25*, 1169–1175. [CrossRef]
120. Karatay, O.; Doğan, M.; Uyar, T.; Çökeliler, D.; Koçum, İ.C. An Alternative Electrospinning Approach With Varying Electric Field for 2-D-Aligned Nanofibers. *IEEE Trans. Nanotechnol.* **2014**, *13*, 101–108. [CrossRef]
121. Fryer, C.; Scharnagl, M.; Helms, C. Electrostatic alignment of electrospun PEO fibers by the gap method increases individual fiber modulus in comparison to non-aligned fibers of similar diameter. *AIP Adv.* **2018**, *8*, 065023. [CrossRef]
122. Cai, X.; Zhu, P.; Lu, X.; Liu, Y.; Lei, T.; Sun, D. Electrospinning of very long and highly aligned fibers. *J. Mater. Sci.* **2017**, *52*, 14004–14010. [CrossRef]
123. Lei, T.; Xu, Z.; Cai, X.; Xu, L.; Sun, D. New Insight into Gap Electrospinning: Toward Meter-long Aligned Nanofibers. *Langmuir* **2018**, *34*, 13788–13793. [CrossRef] [PubMed]
124. Yang, D.; Lu, B.; Zhao, Y.; Jiang, X. Fabrication of aligned fibrous arrays by magnetic electrospinning. *Adv. Mater.* **2007**, *19*, 3702–3706. [CrossRef]
125. Park, S.H.; Yang, D.Y. Fabrication of aligned electrospun nanofibers by inclined gap method. *J. Appl. Polym. Sci.* **2011**, *120*, 1800–1807. [CrossRef]

126. Dabirian, F.; Sarkeshik, S.; Kianiha, A. Production of uniaxially aligned nanofibers using a modified electrospinning method: Rotating jet. *Curr. Nanosci.* **2009**, *5*, 318–323. [CrossRef]
127. Afifi, A.M.; Nakajima, H.; Yamane, H.; Kimura, Y.; Nakano, S. Fabrication of Aligned Poly (L-lactide) Fibers by Electrospinning and Drawing. *Macromol. Mater. Eng.* **2009**, *294*, 658–665. [CrossRef]
128. Li, D.; Wang, Y.; Xia, Y. Electrospinning of polymeric and ceramic nanofibers as uniaxially aligned arrays. *Nano Lett.* **2003**, *3*, 1167–1171. [CrossRef]
129. Jalili, R.; Morshed, M.; Ravandi, S.A.H. Fundamental parameters affecting electrospinning of PAN nanofibers as uniaxially aligned fibers. *J. Appl. Polym. Sci.* **2006**, *101*, 4350–4357. [CrossRef]
130. Katta, P.; Alessandro, M.; Ramsier, R.D.; Chase, G.G. Continuous Electrospinning of Aligned Polymer Nanofibers onto a Wire Drum Collector. *Nano Letters* **2004**, *4*, 2215–2218. [CrossRef]
131. Grasl, C.; Arras, M.M.; Stoiber, M.; Bergmeister, H.; Schima, H. Electrodynamic control of the nanofiber alignment during electrospinning. *Appl. Phys. Lett.* **2013**, *102*, 053111. [CrossRef]
132. Khamforoush, M.; Mahjob, M. Modification of the rotating jet method to generate highly aligned electrospun nanofibers. *Mater. Lett.* **2011**, *65*, 453–455. [CrossRef]
133. Badrossamay, M.R.; McIlwee, H.A.; Goss, J.A.; Parker, K.K. Nanofiber assembly by rotary jet-spinning. *Nano Lett.* **2010**, *10*, 2257–2261. [CrossRef] [PubMed]
134. Yao, J.; Bastiaansen, C.W.; Peijs, T. High strength and high modulus electrospun nanofibers. *Fibers* **2014**, *2*, 158–186. [CrossRef]
135. Edie, D. The effect of processing on the structure and properties of carbon fibers. *Carbon* **1998**, *36*, 345–362. [CrossRef]
136. Dzenis, Y.; Wen, Y. Continuous Carbon Nanofibers For Nanofiber Composites. *MRS Proc.* **2001**, *702*, U5.4.1. [CrossRef]
137. Wang, H.; Wang, H.; Wang, W.; Jin, X.; Lin, T. Research Progress in Polyacrylonitrile (PAN) Based Carbon Nanofibers Electrode Materials for Supercapacitor. *Cailiao Daobao/Mater. Rev.* **2018**, *32*, 730–734, 748. [CrossRef]
138. Eom, Y.; Kim, B.C. Solubility parameter-based analysis of polyacrylonitrile solutions in N, N-dimethyl formamide and dimethyl sulfoxide. *Polymer* **2014**, *55*, 2570–2577. [CrossRef]
139. Chawla, S.; Cai, J.; Naraghi, M. Mechanical tests on individual carbon nanofibers reveals the strong effect of graphitic alignment achieved via precursor hot-drawing. *Carbon* **2017**, *117*, 208–219. [CrossRef]
140. Papkov, D.; Zou, Y.; Andalib, M.N.; Goponenko, A.; Cheng, S.Z.; Dzenis, Y.A. Simultaneously strong and tough ultrafine continuous nanofibers. *Acs Nano* **2013**, *7*, 3324–3331. [CrossRef]
141. Naebe, M.; Lin, T.; Wang, X. *Carbon Nanotubes Reinforced Electrospun Polymer Nanofibres*; Croatia, UK, 2010; Available online: https://www.intechopen.com/books/nanofibers/carbon-nanotubes-reinforced-electrospun-polymer-nanofibres (accessed on 11 December 2020).
142. Chung, D.D. *Carbon Composites: Composites with Carbon Fibers, Nanofibers, and Nanotubes*; Butterworth-Heinemann: Oxford, UK, 2016.
143. Chae, H.G.; Minus, M.L.; Rasheed, A.; Kumar, S. Stabilization and carbonization of gel spun polyacrylonitrile/single wall carbon nanotube composite fibers. *Polymer* **2007**, *48*, 3781–3789. [CrossRef]
144. Papkov, D.; Goponenko, A.; Compton, O.C.; An, Z.; Moravsky, A.; Li, X.Z.; Nguyen, S.T.; Dzenis, Y.A. Improved graphitic structure of continuous carbon nanofibers via graphene oxide templating. *Adv. Funct. Mater.* **2013**, *23*, 5763–5770. [CrossRef]
145. Prilutsky, S.; Zussman, E.; Cohen, Y. The effect of embedded carbon nanotubes on the morphological evolution during the carbonization of poly(acrylonitrile) nanofibers. *Nanotechnology* **2008**, *19*, 165603. [CrossRef] [PubMed]
146. Cai, J.; Naraghi, M. The formation of highly ordered graphitic interphase around embedded CNTs controls the mechanics of ultra-strong carbonized nanofibers. *Acta Mater.* **2019**, *162*, 46–54. [CrossRef]
147. Yu, Y.; Tan, Z.; Zhang, J.; Liu, G. Microstructural evolution and mechanical investigation of hot stretched graphene oxide reinforced polyacrylonitrile nanofiber yarns. *Polym. Adv. Technol.* **2020**. [CrossRef]
148. Peng, K.; Nain, A.; Mirzaeifar, R. Tracking the origins of size dependency in the mechanical properties of polymeric nanofibers at the atomistic scale. *Polymer* **2019**, *175*, 118–128. [CrossRef]
149. Inai, R.; Kotaki, M.; Ramakrishna, S. Structure and properties of electrospun PLLA nanofibers. In Proceedings of the 55th Society of Polymer Science Japan Symposium on Macromolecules, Toyama, Japan, 20–22 September 2006; p. 5507.
150. Wang, D.; Sun, G.; Chiou, B.S.; Hinestroza, J.P. Controllable fabrication and properties of polypropylene nanofibers. *Polym. Eng. Sci.* **2007**, *47*, 1865–1872. [CrossRef]
151. Sehaqui, H.; Ezekiel Mushi, N.; Morimune, S.; Salajkova, M.; Nishino, T.; Berglund, L.A. Cellulose Nanofiber Orientation in Nanopaper and Nanocomposites by Cold Drawing. *ACS Appl. Mater. Interfaces* **2012**, *4*, 1043–1049. [CrossRef]
152. Zhang, Z.; Tu, W.; Peijs, T.; Bastiaansen, C.W.M. Fabrication and properties of poly(tetrafluoroethylene) nanofibres via sea-island spinning. *Polymer* **2017**, *109*, 321–331. [CrossRef]
153. Hooshmand, S.; Aitomäki, Y.; Berglund, L.; Mathew, A.P.; Oksman, K. Enhanced alignment and mechanical properties through the use of hydroxyethyl cellulose in solvent-free native cellulose spun filaments. *Compos. Sci. Technol.* **2017**, *150*, 79–86. [CrossRef]
154. Alarifi, I.M.; Alharbi, A.; Khan, W.S.; Rahman, A.S.; Asmatulu, R. Mechanical and thermal properties of carbonized PAN nanofibers cohesively attached to surface of carbon fiber reinforced composites. In Proceedings of the Macromolecular Symposia, Lincoln, CA, USA, 11–15 May 2015; pp. 140–150.
155. Groover, M.P. *Fundamentals of Modern Manufacturing: Materials Processes, and Systems*; John Wiley & Sons: Hoboken, NJ, USA, 2007.

156. Huang, S.; Zhou, L.; Li, M.-C.; Wu, Q.; Kojima, Y.; Zhou, D. Preparation and properties of electrospun poly (vinyl pyrrolidone)/cellulose nanocrystal/silver nanoparticle composite fibers. *Materials* **2016**, *9*, 523. [CrossRef] [PubMed]
157. Kancheva, M.; Toncheva, A.; Manolova, N.; Rashkov, I. Enhancing the mechanical properties of electrospun polyester mats by heat treatment. *Express Polym. Lett.* **2015**, *9*, 49–65. [CrossRef]
158. Wang, M.; Jin, H.-J.; Kaplan, D.L.; Rutledge, G.C. Mechanical properties of electrospun silk fibers. *Macromolecules* **2004**, *37*, 6856–6864. [CrossRef]
159. Lin, Y.; Clark, D.M.; Yu, X.; Zhong, Z.; Liu, K.; Reneker, D.H. Mechanical properties of polymer nanofibers revealed by interaction with streams of air. *Polymer* **2012**, *53*, 782–790. [CrossRef]
160. Tan, E.; Goh, C.; Sow, C.; Lim, C. Tensile test of a single nanofiber using an atomic force microscope tip. *Appl. Phys. Lett.* **2005**, *86*, 073115. [CrossRef]
161. Khan, W.S.; Asmatulu, R.; Rodriguez, V.; Ceylan, M. Enhancing thermal and ionic conductivities of electrospun PAN and PMMA nanofibers by graphene nanoflake additions for battery-separator applications. *Int. J. Energy Res.* **2014**, *38*, 2044–2051. [CrossRef]
162. Im, J.S.; Kim, J.G.; Bae, T.-S.; Lee, Y.-S. Effect of heat treatment on ZrO2-embedded electrospun carbon fibers used for efficient electromagnetic interference shielding. *J. Phys. Chem. Solids* **2011**, *72*, 1175–1179. [CrossRef]
163. Bhattacharya, M. Polymer nanocomposites—A comparison between carbon nanotubes, graphene, and clay as nanofillers. *Materials* **2016**, *9*, 262. [CrossRef]
164. Lee, K.H.; Kim, H.Y.; Khil, M.S.; Ra, Y.M.; Lee, D.R. Characterization of nano-structured poly(ε-caprolactone) nonwoven mats via electrospinning. *Polymer* **2003**, *44*, 1287–1294. [CrossRef]
165. Wei, Y.; Song, Y.; Deng, X.; Han, B.; Zhang, X.; Shen, Y.; Lin, Y. Dielectric and ferroelectric properties of BaTiO3 nanofibers prepared via electrospinning. *J. Mater. Sci. Technol.* **2014**, *30*, 743–747. [CrossRef]
166. Jabbarnia, A.; Khan, W.; Ghazinezami, A.; Asmatulu, R. Tuning the Ionic and Dielectric Properties of Electrospun Nanocomposite Fibers for Supercapacitor Applications. *Int. J. Eng. Res. Appl.* **2016**, *6*, 65–73.
167. Lee, Y.-I.; Jang, D.-H.; Choa, Y.-H. Synthesis, morphology control and electromagnetic wave absorption properties of electrospun FeCo alloy nanofibers. *J. Nanosci. Nanotechnol.* **2016**, *16*, 5190–5194. [CrossRef] [PubMed]
168. Wang, Z.; Nelson, J.; Koratkar, N.; Schadler, L.; Hillborg, H.; Zhao, S. Dielectric properties of electrospun barium titanate fibers/graphene/silicone rubber composites. In Proceedings of the 2011 Annual Report Conference on Electrical Insulation and Dielectric Phenomena, Cancun, Mexico, 16–19 October 2011; pp. 640–643.
169. Anitha, S.; Natarajan, T. Electrospun Fibrous Nanocomposite Membrane for UV Shielding Applications. *J. Nanosci. Nanotechnol.* **2015**, *15*, 9705–9710. [CrossRef] [PubMed]
170. Kim, S.M.; Kim, S.H.; Choi, M.S.; Lee, J.Y. Electrospun Carbon Nanotube-Reinforced Nanofiber. *J. Nanosci. Nanotechnol.* **2016**, *16*, 2908–2911. [CrossRef] [PubMed]
171. Isaac, B.; Taylor, R.; Adnan, A.; Raihan, R. Electrospinning Approach for the Improvement of Mechanical and Shielding Properties of Nanofiber Mats. In Proceedings of the American Society for Composites—Thirty-Second Technical Conference, West Lafayette, IN, USA, 23–25 October 2017.
172. Yang, D.; Jiang, X. Progress in Ordered Nanofibers via Electrospinning. *Synth. Fiber China* **2008**, 2.
173. Arras, M.M.; Grasl, C.; Bergmeister, H.; Schima, H. Electrospinning of aligned fibers with adjustable orientation using auxiliary electrodes. *Sci. Technol. Adv. Mater.* **2012**, *13*, 035008. [CrossRef]
174. Bashur, C.A. Effect of Electrospun Mesh Diameter, Mesh Alignment, and Mechanical Stretch on Bone Marrow Stromal Cells for Ligament Tissue Engineering. Ph.D. Thesis, Virginia Tech, Blacksburg, VA, USA, 2009.
175. Lawrence, C.; Liu, P. Relation of structure, properties and performance of fibrous media for gas filtration. *Chem. Eng. Technol. Ind. Chem. Plant Equip. Process Eng. Biotechnol.* **2006**, *29*, 957–967. [CrossRef]
176. Katti, D.S.; Robinson, K.W.; Ko, F.K.; Laurencin, C.T. Bioresorbable nanofiber-based systems for wound healing and drug delivery: Optimization of fabrication parameters. *J. Biomed. Mater. Res. Part B* **2004**, *70*, 286–296. [CrossRef]
177. Hou, H.; Ge, J.J.; Zeng, J.; Li, Q.; Reneker, D.H.; Greiner, A.; Cheng, S.Z. Electrospun polyacrylonitrile nanofibers containing a high concentration of well-aligned multiwall carbon nanotubes. *Chem. Mater.* **2005**, *17*, 967–973. [CrossRef]
178. Kannan, P.; Eichhorn, S.J.; Young, R.J. Deformation of isolated single-wall carbon nanotubes in electrospun polymer nanofibres. *Nanotechnology* **2007**, *18*, 235707. [CrossRef]
179. Terao, T.; Zhi, C.; Bando, Y.; Mitome, M.; Tang, C.; Golberg, D. Alignment of boron nitride nanotubes in polymeric composite films for thermal conductivity improvement. *J. Phys. Chem. C* **2010**, *114*, 4340–4344. [CrossRef]
180. Dhakate, S.; Chaudhary, A.; Gupta, A.; Pathak, A.; Singh, B.; Subhedar, K.; Yokozeki, T. Excellent mechanical properties of carbon fiber semi-aligned electrospun carbon nanofiber hybrid polymer composites. *RSC Adv.* **2016**, *6*, 36715–36722. [CrossRef]
181. Moreland, J.C. *Production and Characterization of Aramid Copolymer Fibers for Use in Cut Protection*; ProQuest LLC.: Bethesda, MD, USA, 2010.
182. Zhang, Q.; Wang, Q.; Chen, Y. Structure and tensile properties of melt spun UHMWPE fibers in drawing process. *Gaofenzi Cailiao Kexue Yu Gongcheng/Polym. Mater. Sci. Eng.* **2014**, *30*, 80–84.
183. Picken, S.J.; de Ruijter, C.; Mendes, E.; Boerstoel, H. Orientational order and mechanical properties of poly(amide-block-aramid) alternating block copolymer films and fibers. *Polymer* **2006**, *47*, 8517–8526. [CrossRef]
184. Yu, H.; Potter, K.D.; Wisnom, M.R. A novel manufacturing method for aligned discontinuous fibre composites (High Performance-Discontinuous Fibre method). *Compos. Part A Appl. Sci. Manuf.* **2014**, *65*, 175–185. [CrossRef]

185. Compton, B.G.; Lewis, J.A. 3D-printing of lightweight cellular composites. *Adv. Mater.* **2014**, *26*, 5930–5935. [CrossRef]
186. Malek, S.; Raney, J.R.; Lewis, J.A.; Gibson, L.J. Lightweight 3D cellular composites inspired by balsa. *Bioinspir. Biomim.* **2017**, *12*, 026014. [CrossRef]
187. Ning, N.; Bai, X.; Yang, D.; Zhang, L.; Lu, Y.; Nishi, T.; Tian, M. Dramatically improved dielectric properties of polymer composites by controlling the alignment of carbon nanotubes in matrix. *RSC Adv.* **2014**, *4*, 4543–4551. [CrossRef]
188. Ma, X.; Liu, J.; Ni, C.; Martin, D.C.; Chase, D.B.; Rabolt, J.F. Molecular orientation in electrospun poly (vinylidene fluoride) fibers. *ACS Macro Lett.* **2012**, *1*, 428–431. [CrossRef]
189. Agarwal, S.; Greiner, A.; Wendorff, J.H. Electrospinning of manmade and biopolymer nanofibers—Progress in techniques, materials, and applications. *Adv. Funct. Mater.* **2009**, *19*, 2863–2879. [CrossRef]
190. Edmondson, D.; Cooper, A.; Jana, S.; Wood, D.; Zhang, M. Centrifugal electrospinning of highly aligned polymer nanofibers over a large area. *J. Mater. Chem.* **2012**, *22*, 18646–18652. [CrossRef]
191. Kumar, P.; Kumar, A.; Cho, K.Y.; Das, T.K.; Sudarsan, V. An asymmetric electrically conducting self-aligned graphene/polymer composite thin film for efficient electromagnetic interference shielding. *AIP Adv.* **2017**, *7*, 015103. [CrossRef]
192. Song, W.-L.; Cao, M.-S.; Lu, M.-M.; Yang, J.; Ju, H.-F.; Hou, Z.-L.; Liu, J.; Yuan, J.; Fan, L.-Z. Alignment of graphene sheets in wax composites for electromagnetic interference shielding improvement. *Nanotechnology* **2013**, *24*, 115708. [CrossRef] [PubMed]

MDPI
St. Alban-Anlage 66
4052 Basel
Switzerland
Tel. +41 61 683 77 34
Fax +41 61 302 89 18
www.mdpi.com

Fibers Editorial Office
E-mail: fibers@mdpi.com
www.mdpi.com/journal/fibers

www.ingramcontent.com/pod-product-compliance
Lightning Source LLC
Chambersburg PA
CBHW062106220926
51850CB00010B/2916
* 9 7 8 3 0 3 6 5 2 9 4 3 1 *